FUNDAMENTAL CONCEPTS AND TECHNIQUES OF CALCULUS

Math Refresher Course

REINHARD O.W. FRANZ

Brigham Young University

BYU ACADEMIC PRINTING
Provo, Utah 84602

August 16, 2010

CONTENTS

1

BASIC TECHNIQUES OF ALGEBRA

1.1 Definitions and Results

1.1.1 The Language of Mathematics

To facilitate the clear and concise formulation of mathematical statements, we begin reviewing some extensions to our natural language which will allow us to talk about abstract entities. We will confine ourselves to the basic concepts of "naive set theory" as opposed to "axiomatic set theory".

Although this new terminology and notation might be intimidating at first, the reader will soon realize its usefulness as we move along and he becomes more fluent in using the new terms and notation of this "language".

Figure 1.1: *Georg Cantor (1845–1918)*

Sets and Subsets

According to G. Cantor a *set* is a well-defined collection of distinct "objects", which are called the *elements* or *members* of the set.

Definition 1.1 (Cardinality of a Set)
If S is a set, then the number of its elements of S is called the *cardinality* of S and denoted by $\#(S)$.

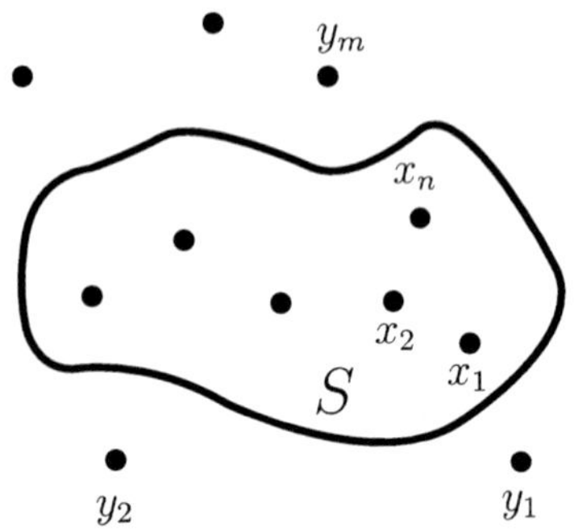

Figure 1.2: *Venn Diagram of a Set.*
The figure shows the "Venn-diagram" of the set S containing the elements $x_1, x_2, \ldots, x_n$. The elements of the set are enclosed by a curve separating the elements belonging to the set from those that do not belong to the set. Here, for instance, the elements $y_1, y_2, \ldots, y_m$ do not belong to the set S.

Often it is more convenient not to represent the elements of the set by dots, but to let the region inside the curve represent the elements of the set, while the region outside represents the elements that do not belong to the set.

Although this is not a valid definition (since it just replaces the name "set" by the equally undefined name "collection"), it it is sufficient for our purpose. For us a set S is well-defined if for each object x it can be decided whether it is an element of S, which we will denote by $x \in S$, or whether it is not an element of S, which will be expressed by $x \notin S$.

We can describe a set S by listing its elements in the following way between "curly" parenthesis: for instance:

$$S = \{x_1, x_2, \ldots, x_n\}, \tag{1.1}$$

which can be expressed by "S is the set consisting of the elements $x_1, x_2, \ldots, x_n$, etc." This is often called the *tabular* form of the set (cf. Figure 1.1.1). Another way for defining a set is by specifying a property P that its elements are to satisfy:

$$S = \{x \mid x \text{ has the property } P\}. \tag{1.2}$$

Thus S is the set of all those elements that satisfy the property P. This way of defining sets is often referred to as the *set-builder* or *descriptive* form. If all elements belong to some bigger set N from which they are selected by the property P, we simply write

$$S = \{x \in N \mid x \text{ has the property } P\}.$$

Definition 1.2 (Empty Set)
The unique set $\emptyset := \{x \mid x \neq x\}$ which does not contain any element is called the *empty set.*

Example 1.1

1. $\mathbb{R} := \{x \mid x \text{ is a real number}\}$
   This is the set of all real numbers, which is usually denoted by the "blackboard bold" version of the letter R. The symbol ":=" ("equal by definition") indicates that the expression on the left (where the colon is) is defined by the expression on the right.

2. $\mathbb{N} := \{n \mid n \text{ is a natural number}\} = \{1, 2, 3, \ldots\}$.

3. $\mathbb{N}^0 := \{n \mid n \text{ is a natural number or } n = 0\} = \{0, 1, 2, 3, \ldots\}$.

4. $\mathbb{Z} := \{z \mid z \text{ is an integer}\} = \{\ldots, -3, -2, -1, 0, 1, 2, 3, \ldots\}$.

5. $\mathbb{Q} := \{x \mid x \text{ is a rational number}\} = \{\frac{p}{q} \mid p \in \mathbb{Z} \text{ and } q \in \mathbb{N}\}$.

6. $\mathbb{C} := \{z \mid z \text{ is a complex number}\} = \{x + yi \mid x, y \in \mathbb{R}\}$.

Definition 1.3 (Intervals)
Let $a, b \in \mathbb{R}$, then we define

1. $[a, b] := \{x \in \mathbb{R} \mid a \leq x \leq b\}$
2. $(a, b) := \{x \in \mathbb{R} \mid a < x < b\}$
3. $[a, b) := \{x \in \mathbb{R} \mid a \leq x < b\}$
4. $(a, b] := \{x \in \mathbb{R} \mid a < x \leq b\}$

and call the sets $[a, b]$, (a, b), $[a, b)$ and $(a, b]$ the *closed*, *open*, and *half-open intervals* defined by a and b. Note that these intervals are empty if $a > b$.

We will now consider two fundamental relationships between sets, "equality" and the "subset" relationship.

Definition 1.4 (Equality of Sets)
The sets A and B are *equal*, which we abbreviate by $A = B$, if they possess the same elements, i.e.: if for all x holds: $x \in A \Leftrightarrow x \in B$. If the sets A and B are not equal, we write $A \neq B$.

Recall that for propositions α and β the expression "$\alpha \Rightarrow \beta$" is a shorthand for "α implies β" or "if α then β". The implication "$\alpha \Rightarrow \beta$" is only false if α is true and β is false. In particular, "$\alpha \Rightarrow \beta$" is true if α is false.

"$\alpha \Leftrightarrow \beta$" is short for "$\alpha$ and β are equivalant", i.e. α is true if and only if β is true. Hence, $\alpha \Leftrightarrow \beta$ is identical to the conjunction of the two propositions: "$\alpha \Rightarrow \beta$ and $\beta \Rightarrow \alpha$". See the truth table for the implication in the margin.

Example 1.2

1. $\mathbb{N} \neq \mathbb{N}^0$, since $0 \in \mathbb{N}^0$, however $0 \notin \mathbb{N}$.
2. $\{1, 2, 3, 4\} = \{1, 1, 1, 4, 2, 3, 3\}$.

Definition 1.5 (Subsets)
A set A is called a *subset* of a set B, which we abbreviate by $A \subseteq B$, if for all x holds: $x \in A \Rightarrow x \in B$. If A is not a subset of B we write $A \not\subseteq B$.

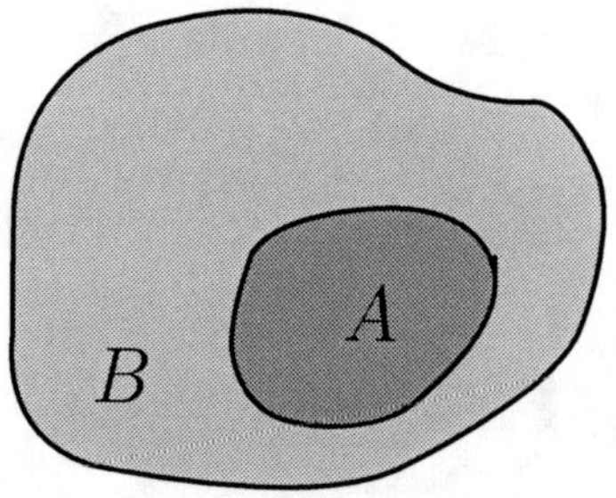

Figure 1.3: *Subset*
The set A, represented by the dark region, is a subset of the set B, represented by the light region containing the dark region.

Example 1.3

1. $S \subseteq S$, since $x \in S \Rightarrow x \in S$.
2. $\emptyset \subseteq S$ for any set S, since "$x \in \emptyset \Rightarrow x \in S$" is true as, by definition, "$x \in \emptyset$" is false for any element x.
3. $\mathbb{N} \subseteq \mathbb{N}^0$, $\mathbb{N}^0 \subseteq \mathbb{Z}$, $\mathbb{Z} \subseteq \mathbb{Q}$, $\mathbb{Q} \subseteq \mathbb{R}$, $\mathbb{R} \subseteq \mathbb{C}$.
4. $\{1, 2, 3, 4, 5\} \not\subseteq \{2, 3, 4, 5, 6\}$.
5. $\{2, 3, 4, 5, 6\} \not\subseteq \{1, 2, 3, 4, 5\}$.

The following useful statements concerning the inclusion "$\subseteq$" can be easily established. We list them as

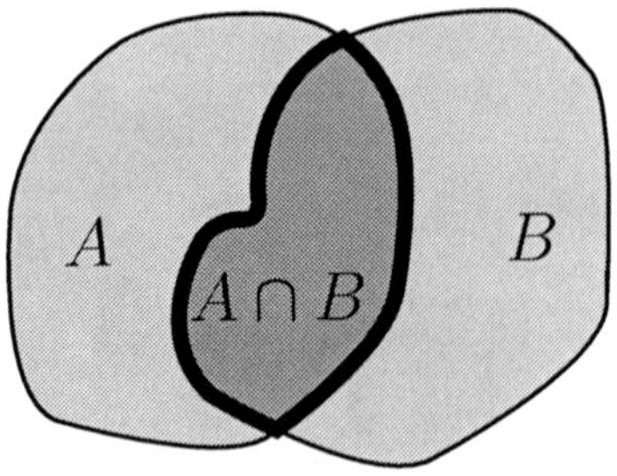

Figure 1.4: *Intersection Set.*
The intersection $A \cap B$ of the sets A and B is represented by the dark region common to both light regions representing the sets A and B.

Lemma 1.1 (Properties of "$\subseteq$")
If M, N, and L are sets then the following statements hold:

1. *If $M \subseteq N$ and $N \subseteq M$ then $M = N$.* (anti-symmetric)
2. *If $M \subseteq N$ and $N \subseteq L$ then $M \subseteq L$.* (transitive)

To verify the correctness of these statements, we have to show that they follow from the definitions of "$=$" and "$\subseteq$", respectively. We include the simple proof as an easy example for mathematical reasoning.

PROOF: (1) Suppose $M \subseteq N$ and $N \subseteq M$. We have to show that M and N possess the same elements. However, $M \subseteq N$ implies by definition that "$x \in M \Rightarrow x \in N$", and $N \subseteq M$ implies similarly that "$x \in N \Rightarrow x \in M$". Hence, $x \in M \Leftrightarrow x \in N$ and thus, by definition, $M = N$.

(2) Now suppose that $M \subseteq N$ and $N \subseteq L$. We have to show that $x \in M \Rightarrow x \in L$. However, $M \subseteq N$ implies $x \in M \Rightarrow x \in N$ and $N \subseteq L$ implies $x \in N \Rightarrow x \in L$. Hence $x \in M \Rightarrow x \in L$, which means $M \subseteq L$. $\square$

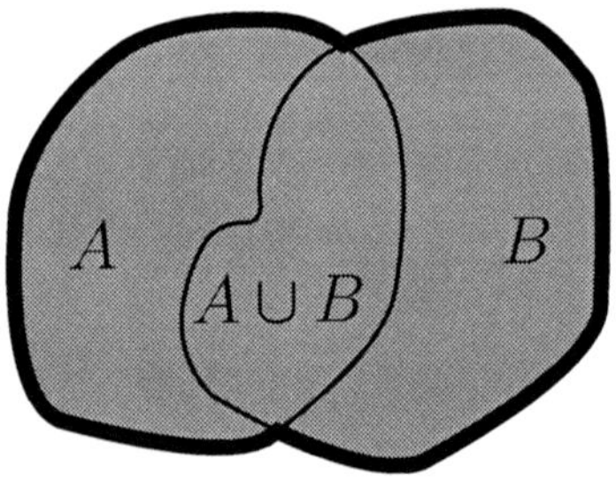

Figure 1.5: *Union Set.*
The union $A \cup B$ of the sets A and B is represented by the dark region comprising the regions A and B.

Set Operations

We will now describe operations which allow the construction of new sets from old sets.

Definition 1.6 (Intersection)
For sets A and B the set

$$A \cap B := \{x \mid x \in A \text{ and } x \in B\}$$

is called the *intersection* of A and B. If $A \cap B = \emptyset$, the sets A and B are called *disjoint*.

Definition 1.7 (Union)
If A and B are sets then the set

$$A \cup B := \{x \mid x \in A \text{ or } x \in B\}$$

is called the *union* of A and B.

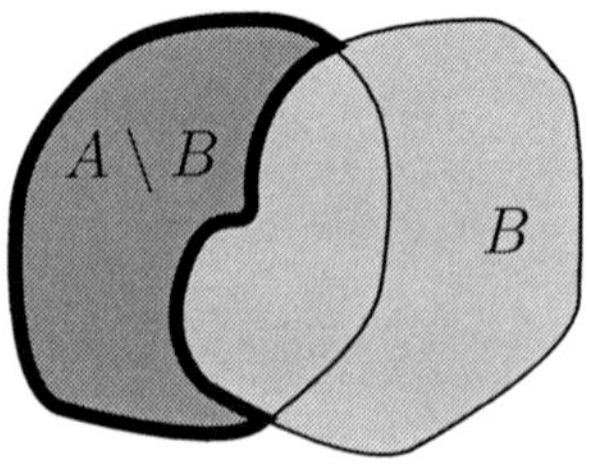

Figure 1.6: *Difference Set.*
The difference $A \setminus B$ of the sets A and B is represented by the dark region consisting of that part of the region A that does not intersect with the region B.

Definition 1.8 (Difference)
If A and B are sets then the set

$$A \setminus B := \{x \mid x \in A \text{ and } x \notin B\}$$

is called the *difference* of A and B. $A \setminus B$ is usually read as "A without B".

Definition 1.9 (Cartesian Product)
If A and B are sets then the set

$$A \times B := \{(x, y) \mid x \in A \text{ and } y \in B\}$$

is called the *Cartesian* product of the sets A and B. The elements (x, y) of $A \times B$ are called *ordered pairs.* Two ordered pairs (x, y) and (x', y') are defined to be equal if $x = x'$ and $y = y'$. If $A = B$, we often write A^2 instead of $A \times A$. Similarly, if $A_1, \ldots, A_n$ are sets, we denote by by

$$A_1 \times \cdots \times A_n := \prod_{k=1}^{n} A_k := \{(a_1, \ldots, a_n) \mid a_k \in A_k,\ k = 1, \ldots, n\}$$

the Cartesian product of the sets $A_1 \ldots, A_n$, whose elements $(a_1, \ldots, a_n)$ are called n*-tuples.* Again, if $A_1 = \cdots = A_n = A$, we usually denote their Cartesian $A \times \cdots \times A$ product simply by A^n.

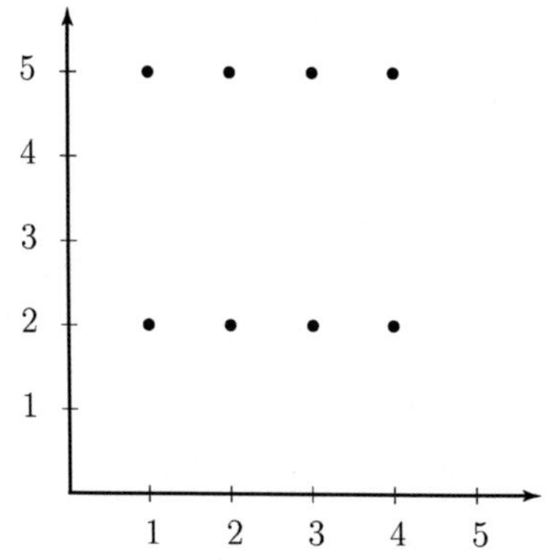

Figure 1.7: *Product Set.*
A visualization of the product set $\{1, 2, 3, 4\} \times \{2, 5\}$

Example 1.4
Let $A := \{1, 2, 3, 4\}$ and $B := \{2, 5\}$ then

$$\begin{aligned} A \cap B &= \{2\}, \\ A \cup B &= \{1, 2, 3, 4, 5\}, \\ A \setminus B &= \{1, 3, 4\}, \\ A \times B &= \{(1,2), (2,2), (3,2), (4,2), (1,5), (2,5), (3,5), (4,5)\} \end{aligned}$$

and

$$B \times A = \{(2,1), (2,2), (2,3), (2,4), (5,1), (5,2), (5,3), (5,4)\}$$

Remark 1.1
Note that the Cartesian product, in general, is *not* commutative, as the previous example illustrates.

Example 1.5
Let $A := (1, 5]$ and $B := [2, 5]$, then

$$A \times B = (1, 5] \times [2, 5] = \{(x, y) \mid 1 < x \leq 5 \text{ and } 2 \leq y \leq 5\}$$

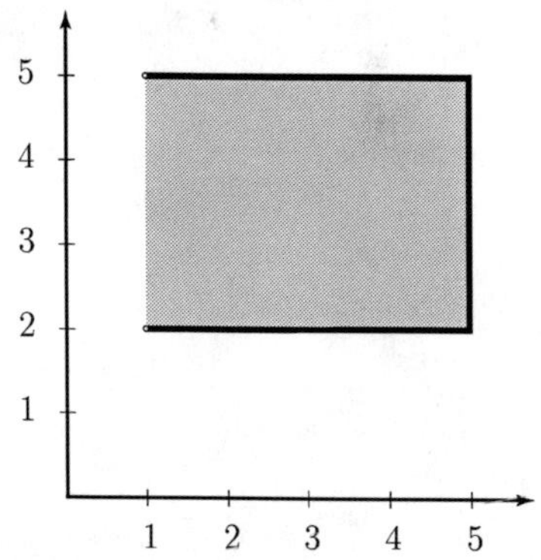

Figure 1.8: *Product Set.*
A visualization of the set $(1, 5] \times (2, 5]$. The bolded edges of the rectangle indicate that part of the boundary of $(1, 5] \times (2, 5]$ that belongs to the product set

Example 1.6
The sets

$$\begin{aligned} \mathbb{R}^2 &= \{(x, y) \mid x, y \in \mathbb{R}\}, \\ \mathbb{R}^3 &= \{(x, y, z) \mid x, y, z \in \mathbb{R}\}, \end{aligned}$$

or more general,

$$\mathbb{R}^n = \{(x_1, x_2, \ldots, x_n) \mid x_1, \ldots, x_n \in \mathbb{R}\}$$

represent the Euclidean plane, the 3-dimensional Euclidean space and the n-dimensional Euclidean space, respectively.

Some important properties of the set operations introduced above are listed for easy reference in the following

Lemma 1.2 (Algebraic Properties of "$\cap$", "$\cup$" and "$\setminus$")
For sets M, N, and L the following properties hold:

1. $M \cap N = N \cap M$ and $M \cup N = N \cup M$ *(commutativity)*
2. $M \cap (N \cap L) = (M \cap N) \cap L$ *(associativity)*
   $M \cup (N \cup L) = (M \cup N) \cup L$
3. $M \cap (N \cup L) = (M \cap N) \cup (M \cap L)$ *(distributive law)*
   $M \cup (N \cap L) = (M \cup N) \cap (M \cup L)$
4. $M \setminus (N \cap L) = (M \setminus N) \cup (M \setminus L)$ *(De Morgan's rules)*
   $M \setminus (N \cup L) = (M \setminus N) \cap (M \setminus L)$

Suppose S is a set and P a statement concerning the elements x of S, then it is often convenient to abbreviate the lengthy formulation

"for all $x \in S$ the statement P is true"

by the more compact symbolic expression

$$\forall x \in S : P.$$

Similarly, we will often replace the phrase

"there exists (at least) one $x \in S$ for which the statement P is true"

by the shorter expression

$$\exists x \in S : P.$$

Example 1.7
Suppose A and B are sets then, by the definition of "$\subseteq$" and "$\nsubseteq$", we have

$$A \subseteq B \Leftrightarrow (\forall x \in A : x \in B)$$
$$A \nsubseteq B \Leftrightarrow (\exists x \in A : x \notin B)$$

Example 1.8
The lengthy formulation of the "Archimidean Axiom": "For each real number x there exists a natural number n which is greater than or equal to x" can be shortened to

$$\forall x \in \mathbb{R} \ \exists n \in \mathbb{N} : n \geq x$$

Definition 1.10 (Generalized Intersection and Union)
Suppose I is a nonempty set and for each $i \in I$ let S_i be a set.

1. The *(generalized) intersection* of the sets S_i is defined as the set

$$\bigcap_{i \in I} S_i := \{x \mid \forall i \in I : x \in S_i\}$$

2. The *(generalized) union* of the sets S_i is defined as the set

$$\bigcup_{i \in I} S_i := \{x \mid \exists i \in I : x \in S_i\}$$

If $I = \{1, \ldots, n\}$ we write $\bigcap_{i=1}^{n} S_i$ instead of $\bigcap_{i \in \{1,\ldots,n\}} S_i$ and similarly for the union.

Lemma 1.3 (De Morgan's Rules)
Suppose S is a set, I a nonempty set and for each $i \in I$ let S_i be a set. Then holds

1. $S \setminus \left(\bigcap_{i \in I} S_i \right) = \bigcup_{i \in I} (S \setminus S_i)$.

2. $S \setminus \left(\bigcup_{i \in I} S_i \right) = \bigcap_{i \in I} (S \setminus S_i)$.

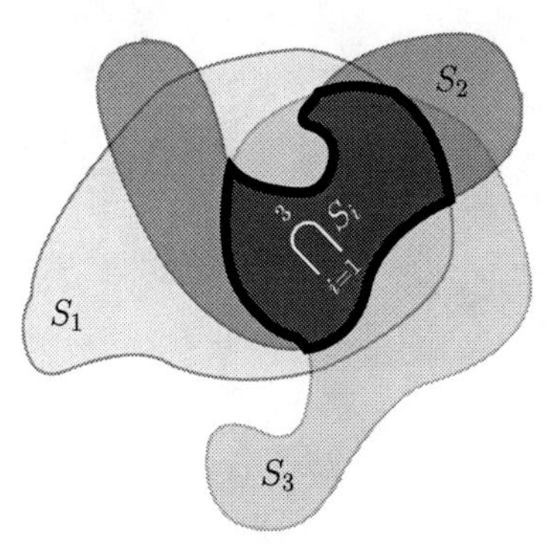

Figure 1.9: *Generalized Intersection.*
The intersection $\bigcap_{i=1}^{3} S_i = S_1 \cap S_2 \cap S_3$ of the sets S_1, S_2, S_2.

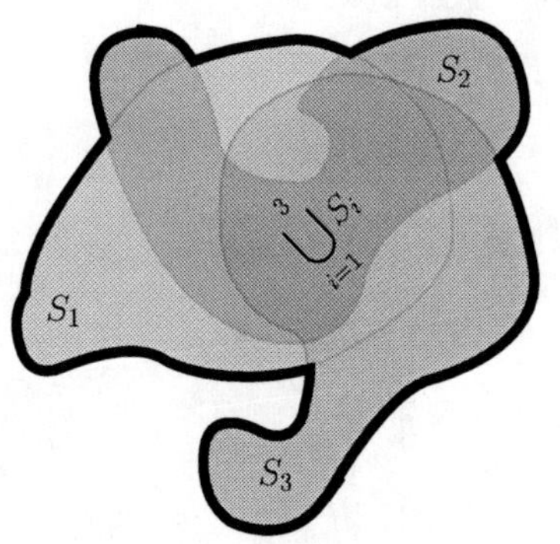

Figure 1.10: *Generalized Union.*
The union $\bigcup_{i=1}^{3} S_i = S_1 \cap S_2 \cap S_3$ of the sets S_1, S_2, S_2.

Functions

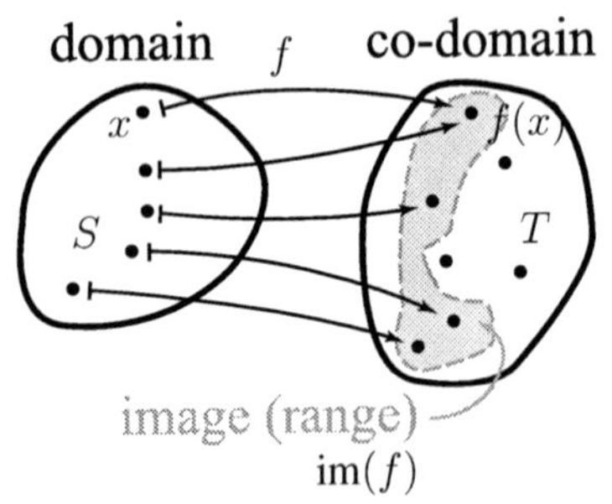

Figure 1.11: *Function.*
The function $f : S \to T$ maps the elements of its domain S into its co-domain T. The set $\mathrm{im}(f)$ of all images of f (also called the range of f, denoted $\mathrm{range}(f)$) is the shaded subset of T.

Definition 1.11 (Function, Domain, Range, Image)

1. Suppose S and T are sets. A *function* or *map* from S into T is an association which assigns to every element of S precisely one element of T. Instead of the somewhat lengthy formulation "f is a function from S into T", we shall often write $f : S \to T$.

2. If $f : S \to T$ is a function and $x \in S$ then we denote the unique element assigned to x by $f(x)$ and call it the *value* of f at x or the *image* of x under f.

3. If $f : S \to T$ is a function, then the set S is called the *domain* of f, the set T the *co-domain* of f, and the set
$$\mathrm{im}(f) = \{f(x) \mid x \in S\}$$
of all images of S under f the *image* of f. or the *range* of f. The image of a function f is often also called the *range* of f and denoted by $\mathrm{range}(f)$.

4. Moreover, if f is a function from S into T, it is often convenient to use the notation $x \mapsto f(x)$ to indicate that $f(x)$ is assigned to x. Note that the arrow "$\to$" describes the global association between the sets S and T, while "$\mapsto$" describes the local association between the elements of S and T.

5. Two functions $f_1 : S_1 \to T_1$ and $f_2 : S_2 \to T_2$ are *equal* if $S_1 = S_2$, $T_1 = T_2$ and
$$f_1(x) = f_2(x)$$
for all $x \in S_1 = S_2$.

Remark 1.2
In applications, real-valued functions of one or several real variables are often defined (sloppily) by an equation of the form
$$y = f(x_1, \ldots, x_n)$$
without explicite indication of their domain and codomain. For instance,
$$y = \frac{x-2}{\sqrt{1-x^2}}.$$
This is done with the understanding that the domain is the largest subset of $\mathbb{R}^n$ for which the term $f(x_1, \ldots, x_n)$ is defined, and that the codomain is $\mathbb{R}$.

Example 1.9
Let $S = T = \mathbb{R}$, and let $f : \mathbb{R} \to \mathbb{R}$ be the map defined by $f(x) := x^2$ for

all $x \in \mathbb{R}$. We can also express this by saying that f is the map from $\mathbb{R}$ into $\mathbb{R}$ such that $x \mapsto x^2$ for all $x \in \mathbb{R}$. The image of f is the interval $[0, \infty)$.

Example 1.10
Find the domain and range of the function

$$f(x) = \frac{\sqrt{9 - x^2}}{x^2 - 3x + 2}.$$

SOLUTION: (i) Domain of f: The domain of f is the largest subset of $\mathbb{R}$ for which

$$f(x) = \frac{\sqrt{9 - x^2}}{x^2 - 3x + 2} = \frac{\sqrt{9 - x^2}}{(x - 2)(x - 1)} = \frac{\sqrt{9 - x^2}}{\left(x - \frac{3}{2}\right)^2 - \frac{1}{4}}$$

is defined. Therefore, we have to eliminate those $x \in \mathbb{R}$ for which $0 = x^2 - 3x + 2 = (x - 1)(x - 2)$, i.e. $x = 1$ and $x = 2$, and confine ourselves to those $x \in \mathbb{R}$ for which $0 \leq 9 - x^2 = (3 - x)(3 + x)$, i.e. $x \in [-3, 3]$. Thus, the domain of f is given by

$$D_f = [-3, 3] \setminus \{1, 2\}.$$

Since $f(1) \not{0}$ and $f(2) \neq 0$, f has vertical tangents at 1 and 2 which are also called poles of f. Since the multiplicity of the linear factors $(x-1)$ and $(x-2)$ are both 1, each pol is of order 1.

(ii) Range of f: To determine the range of f, we will first compute the range of its restrictions to each of the subdomains given by the intervals $[-3, 1)$, $(1, 2)$ and $(2, 3]$. Note that the *only* zeroes of f are the boundary points -3, and 3 and also note that f is continuous on its domain. We can therefore apply the Intermediate-value Theorem to each closed subinterval of its domain:

We begin with the subinterval $[-3, 1)$. Since $f(-3) = 0$, $f(x) \geq 0$ for all $x \in [-3, 1)$ and $\lim_{x \to 1^-} f(x) = \infty$, as can be easily verified, we conclude using the Intermediate-value Theorem that the range of the restriction of f to the interval $[-3, 1)$ is $[0, \infty)$.

We will now consider the behavior of f on the subinterval $(1, 2)$. Clearly, both $\lim_{x \to 1^+} f(x) = \lim_{x \to 2^-} f(x) = -\infty$ and $f(x) < 0$ for all $x \in (1, 2)$. It can be easily verified that f attains its global maximum on the subinterval $(1, 2)$ where the denominator of f which is given by the parabola $\left(x - \frac{3}{2}\right)^2 - \frac{1}{4}$ has its smallest value which is precisely at $\frac{3}{2}$ where the parabola has its vertex. The corresponding function value of f is

$$f\left(\tfrac{3}{2}\right) = \frac{\sqrt{9 - \left(\frac{3}{2}\right)^2}}{-\frac{1}{4}} = -6\sqrt{3}.$$

Hence, again by the Intermediate-value Theorem, the range of the restriction of f to the interval $(1, 2)$ is given by the interval $[-6\sqrt{3}, -\infty)$.

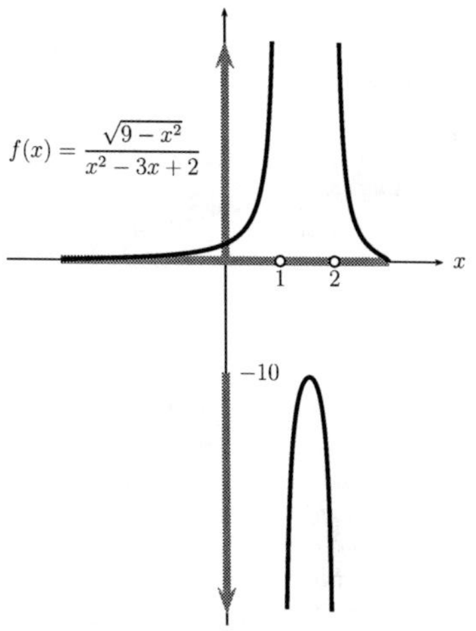

Figure 1.12: *Domain and Range.* The graph of the function f. Note that f has poles at 1 and 2. The domain $D_f = [-3, 3]$ of f and its range (or image) range$(f) = [-6\sqrt{3}, -\infty) \cup [0, \infty)$ are are shaded.

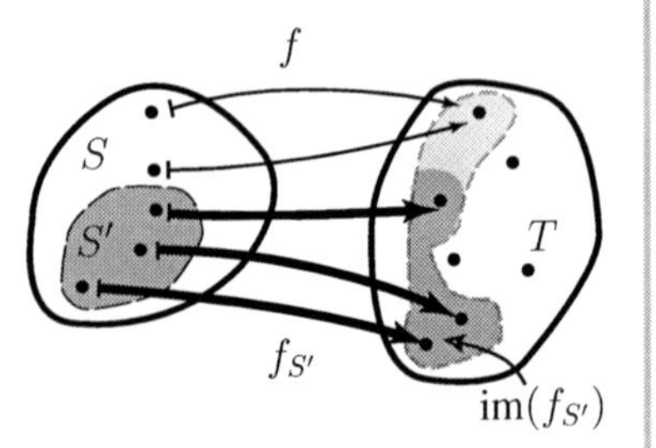

Figure 1.13: *Restriction.*
The figure depicts the restriction $f_{S'}$ of the function $f : S \to T$. The function $f_{S'}$ assigns to each element $x \in S'$ the element $f(x)$ as image. $f_{S'}$ has therefore domain S', and co-domain T. Domain and image of $f_{S'}$ are shaded dark-gray in the illustration. Note that $\mathrm{im}(f_{S'}) \subseteq \mathrm{im}(f)$

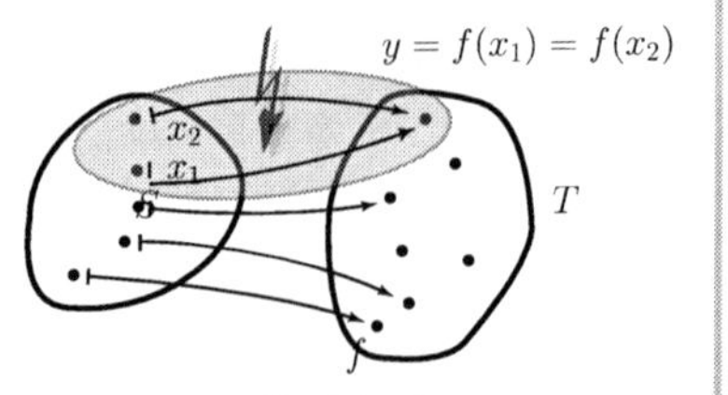

Figure 1.14: *Not Injective.*
The function $f : S \to T$ is *not* injective since the elements x_1 and x_2 are mapped on the same image $y = f(x_1) = f(x_2)$.

Finally, consider the interval $(2, 3]$. Clearly, $f(x) \geq 0$ for all $x \in (2, 3]$, $f(3) = 0$ and $\lim_{x \to 2^+} = \infty$, as can be easily verified. Hence, again by the Intermediate-value Theorem, the range of the restriction of f to the interval $(2, 3]$ is $[0, \infty)$.

The range of f on its total domain $D_f = [-3, 1) \cup (1, 2) \cup (2, 3]$ is therefore the union of the ranges of the restrictions of f. Hence

$$\begin{aligned} \mathrm{range}(f) &= [0, \infty) \cup [-6\sqrt{3}, -\infty) \cup [0, \infty) \\ &= [-6\sqrt{3}, -\infty) \cup [0, \infty). \end{aligned}$$

$\triangleleft$

Definition 1.12 (Restriction of a Function)
If $f : S \to T$ is a map and $S' \subseteq S$ a subset of S, then we can define another map from S' into T by the same association $x \mapsto f(x)$ for all $x \in S'$. This map is called the *restriction* of f to S' and is denoted by $f|_{S'} : S' \to T$.

Definition 1.13 (Injective)
A map $f : S \to T$ is called *injective* or *one-to-one* if for all $x_1, x_2 \in S$ whenever $x_1 \neq x_2$ then also $f(x_1) \neq f(x_2)$. Note this is equivalent to the statement

$$\forall x_1 \forall x_2 \Big(f(x_1) = f(x_2) \Rightarrow x_1 = x_2 \Big).$$

Remark 1.3
Recall the following geometric formulation of Definition 1.13, called the *horizontal line test:* Suppose $S, T \subseteq \mathbb{R}$. Then the function $f : S \to T$ is injective, if any horizontal line (line parallel to the x-axis) intersects the graph of f in at most one point.

Example 1.11
The map $f(x) := x^2$ for all $x \in \mathbb{R}$ defined in the previous example is *not* injective, since, for example $f(-1) = f(1)$. However, the map $g : \mathbb{R} \to \mathbb{R}$ defined by $g(x) := x + 1$ is injective, since $x_1 + 1 = x_2 + 1$ implies $x_1 = x_2$ for all $x_1, x_2 \in \mathbb{R}$.

Definition 1.14 (Surjective)
A map $f : S \to T$ is called *surjective* or *onto* if its image $f(S) = T$, i.e. if for any $y \in T$ there exists an $x \in S$ such that $f(x) = y$. Those elements $x \in S$ with $f(x) = y$ are called *inverse images* of y.

Example 1.12
The map $f(x) := x^2$ for all $x \in \mathbb{R}$ defined above is *not* surjective, since, for example -1 does not have an inverse image. However, the map $g : \mathbb{R} \to \mathbb{R}$ defined by $g(x) := x+1$ is surjective, since for all $y \in \mathbb{R}$ we have $y = g(y-1)$.

Definition 1.15 (Bijective)
A map $f : S \to T$ is called *bijective* if it is both injective and surjective.

Definition 1.16 (Identity Map)
Let S be a nonempty set, then the map $\mathrm{id} = \mathrm{id}_S : S \to S$ defined by $x \mapsto x$ is called the *identity map* on S.

Remark 1.4
Note that id is both injective and surjective, hence bijective.

Definition 1.17 (Composition)
Suppose S, T and U are sets and $f : S \to T$ and $g : T \to U$ are maps. Then we define the *composition* $g \circ f : S \to U$ of f and g by

$$(g \circ f)(x) := g\big(f(x)\big)$$

for all $x \in S$.

Example 1.13 ("$\circ$" is *not* commutative)
Consider again the maps $f, g : \mathbb{R} \to \mathbb{R}$ defined by $f(x) := x^2$ and $g(x) := x + 1$ for all $x \in \mathbb{R}$. Then we can form both $g \circ f$ and $f \circ g$ and we obtain

$$g\big(f(x)\big)(x) = g(x^2) = x^2 + 1,$$

while

$$f\big(g(x)\big)(x) = f(x+1) = (x+1)^2$$

for all $x \in \mathbb{R}$. Hence, we see that in general

$$f \circ g \neq g \circ f.$$

Example 1.14
Suppose that

$$f(x) = \ln(1 - x)$$

and

$$g(x) = x^2.$$

Determine the domain of the composition $f \circ g$ of f and g.

SOLUTION: Since ln is only defined for $x > 0$, the function f has the domain $D_f = (-\infty, 1)$. g is defined on $\mathbb{R}$. Thus, the domain of $f \circ g$ can only contain those $x \in \mathbb{R}$ for which $g(x) = x^2 \in (-\infty, 1)$, ie. for which $x^2 < 1$. Hence, the domain of $f \circ g$ is given by

$$D_{f \circ g} = (-1, 1).$$

◁

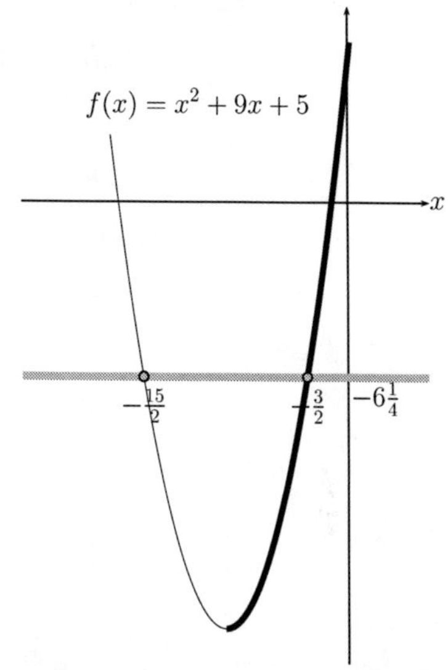

Figure 1.15: *Horizontal Line Test.*
The function $f : \mathbb{R} \to \mathbb{R}$ defined by $f(x) = x^2 + 9x + 5$ is *not* injective, since (for instance) the horizontal line through $-6\frac{1}{4}$ intersects the graph of f in *two* points, namely at $x_1 = -\frac{15}{2}$ and $x_2 = -\frac{3}{2}$. However, f can be made injective by restricting its domain (for instance) to the interval $[-\frac{9}{2}, \infty)$. The graph of the restriction of f to this interval is bolded.

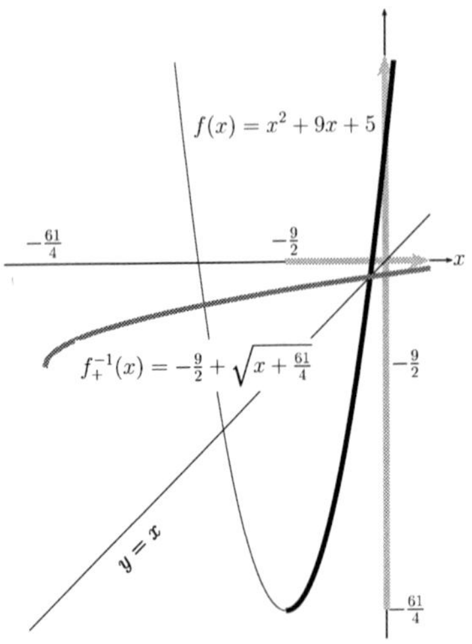

Figure 1.16: *Inverse Function.*
The graph of the restriction f_+ of the function f defined by $f(x) = x^2 + 9x + 5$ is depicted with a bolded curve, its domain $[-\frac{9}{2}, \infty)$ and range $[-\frac{61}{4}, \infty)$ are shaded. The graph of its inverse f_+^{-1} is depicted with a darkly shaded curve. Note that the domain of f_+^{-1} coincides with the range of f_+ and the range of f_+^{-1} with the domain of f_+.

Definition 1.18 (Invertible Functions)
The map $f : S \to T$ is called *invertible* if there exists a map $g : T \to S$ such that

$$g \circ f = \mathrm{id}_S \quad \text{and} \quad f \circ g = \mathrm{id}_T . \tag{1.3}$$

If f is invertible then there exists precisely one map g satisfying the equation (1.3). This map is called the *inverse* map of f and usually denoted by f^{-1}.

Lemma 1.4 (Invertiblity Criterion)
If $f : S \to \mathrm{im}(S)$ is invertible if and only if f is injective.

Lemma 1.5
If $f : S \to T$ is either (strictly) increasing or decreasing, then f is invertible.

Example 1.15
Let $f(x) = x^2 + 9x + 5$. Determine whether f has an inverse. If so, find it.

SOLUTION:
Clearly, the function $f : \mathbb{R} \to \mathbb{R}$ defined by

$$f(x) = x^2 + 9x + 5 = \left(x + \tfrac{9}{2}\right)^2 - \tfrac{61}{4}$$

(note that its graph is a parabola!) is not injective (one-to-one), since, for example,

$$\begin{aligned} f\left(\tfrac{1}{2}\right) &= \left(\tfrac{1}{2} + \tfrac{9}{2}\right)^2 - \tfrac{61}{4} \\ &= 5^2 - \tfrac{61}{4} = (-5)^2 - \tfrac{61}{4} \\ &= \left(-\tfrac{19}{2} + \tfrac{9}{2}\right)^2 - \tfrac{61}{4} = f\left(-\tfrac{19}{2}\right). \end{aligned}$$

Therefore, f is not invertible on it "naturual" domain $\mathbb{R}$. However, we can restrict f (for instance) to the subinterval $[-\frac{9}{2}, \infty)$. This restriction, which we will denote by f_+, constitutes a bijection $f_+ : [-\frac{9}{2}, \infty) \to [-\frac{61}{4}, \infty)$, as can be easily verified.

We will now determine the inverse of f_+: To this end, first set $y = \left(x + \frac{9}{2}\right)^2 - \frac{61}{4}$. Next interchange x and y, which leads to the equation $x = \left(y + \frac{9}{2}\right)^2 - \frac{61}{4}$. Then solve for y: Taking the square root of

$$\left(y + \tfrac{9}{2}\right)^2 = x + \tfrac{61}{4}$$

yields

$$\left|y + \tfrac{9}{2}\right| = \sqrt{x + \tfrac{61}{4}}$$

and

$$y + \tfrac{9}{2} = \sqrt{x + \tfrac{61}{4}}$$

since $y \in [-\frac{9}{2}, \infty)$, implies $\left|y + \frac{9}{2}\right| = y + \frac{9}{2}$. Thus

$$y = -\tfrac{9}{2} + \sqrt{x + \tfrac{61}{4}}.$$

Hence, the inverse function

$$f_+^{-1} : [-\tfrac{61}{4}, \infty) \to [-\tfrac{9}{2}, \infty)$$

is given by

$$f_+^{-1}(x) = -\tfrac{9}{2} + \sqrt{x + \tfrac{61}{4}}$$

for $x \in [-\frac{61}{4}, \infty)$.

We conclude by showing that f_+^{-1} is indeed inverse to f on $[-\frac{9}{2}, \infty)$: Let $x \in [-\frac{9}{2}, \infty)$ then

$$\begin{aligned} f_+^{-1}(f_+(x)) &= f_+^{-1}\Big(\big(x + \tfrac{9}{2}\big)^2 - \tfrac{61}{4}\Big) \\ &= -\tfrac{9}{2} + \sqrt{\big(x + \tfrac{9}{2}\big)^2 - \tfrac{61}{4} + \tfrac{61}{4}} \\ &= -\tfrac{9}{2} + \left|x + \tfrac{9}{2}\right| \\ &= -\tfrac{9}{2} + x + \tfrac{9}{2} \\ &= x. \end{aligned}$$

Conversely, let $x \in [-\frac{61}{4}, \infty)$ then

$$\begin{aligned} f_+(f^{-1}(x)) &= f_+\Big(-\tfrac{9}{2} + \sqrt{x + \tfrac{61}{4}}\Big) \\ &= \Big(-\tfrac{9}{2} + \sqrt{x + \tfrac{61}{4}} + \tfrac{9}{2}\Big)^2 - \tfrac{61}{4} \\ &= x + \tfrac{61}{4} - \tfrac{61}{4} \\ &= x. \end{aligned}$$

◁

1.1.2 Basic Operations with Rational Numbers

Definition 1.19 (Equality, Cancellation)
The rational numbers (fractions) $a/b, c/d \in \mathbb{Q}$ are *equal* if $ad = bc$. In particular,

$$\frac{a \cdot k}{b \cdot k} = \frac{a \cdot \not{k}}{b \cdot \not{k}} = \frac{a}{b} \tag{1.4}$$

for all integers $k \in \mathbb{Z}$. The integer $n \in \mathbb{Z}$ is identified with the fraction $n/1$:

$$n = \frac{n}{1} \tag{1.5}$$

for all $n \in \mathbb{Z}$. Note that $a/b = 0$ if and only if $a = 0$.

Example 1.16

$$\frac{1}{2} = \frac{2}{4} = \frac{3}{6} = \frac{4}{8} = \cdots = \frac{1 \cdot k}{2 \cdot k}, \qquad 3 = \frac{3}{1} = \frac{6}{2} = \frac{9}{3} = \cdots \frac{3 \cdot k}{k}$$

for all $k \in \mathbb{Z}$.

Definition 1.20 (Multiplication, Inverse)
Let $a/b, c/d \in \mathbb{Q}$. Then their *product* $(a/b) \cdot (c/d)$ is defined by

$$\frac{a}{b} \cdot \frac{c}{d} := \frac{ac}{bd} \tag{1.6}$$

Note, that if $a \neq 0$ then its (multiplicative) *inverse* is given by

$$\left(\frac{a}{b}\right)^{-1} = \frac{1}{\frac{a}{b}} = \frac{b}{a} \tag{1.7}$$

Hence, since $x \div y := x \cdot y^{-1}$ for $y \neq 0$,

$$\frac{a}{b} \div \frac{c}{d} = \frac{a}{b} \cdot \left(\frac{c}{d}\right)^{-1} = \frac{a \cdot d}{b \cdot c} \tag{1.8}$$

provided $c \neq 0$.

Example 1.17

1. $2^{-1} = \left(\frac{2}{1}\right)^{-1} = \frac{1}{2}$
2. $\left(\frac{1}{2}\right)^{-1} = \frac{2}{1} = 2$
3. $3 \cdot \frac{5}{6} = \frac{3}{1} \cdot \frac{5}{6} = \frac{3 \cdot 5}{1 \cdot 6} = \frac{5}{2}$
4. $\frac{\frac{2}{3}}{\frac{5}{6}} = \frac{2}{3} \cdot \frac{6}{5} = \frac{4}{5}$
5. $\frac{\frac{2}{3}}{4} = \frac{\frac{2}{3}}{\frac{4}{1}} = \frac{2}{3} \cdot \frac{1}{4} = \frac{1}{6}$

Definition 1.21 (Addition, Inverse)
For rational numbers $a/b,\ c/b \in \mathbb{Q}$, their *sum* is defined by

$$\frac{a}{b} + \frac{c}{b} := \frac{a+c}{b} \tag{1.9}$$

or more general

$$\frac{a}{b} + \frac{c}{d} := \frac{ad}{bd} + \frac{bc}{bd} = \frac{ad+bc}{bd} \tag{1.10}$$

for all $a/b,\ c/d \in \mathbb{Q}$. The (additive) *inverse* of $a/b \in \mathbb{Q}$ is

$$-\frac{a}{b} = \frac{-a}{b} = \frac{a}{-b} \tag{1.11}$$

hence

$$\frac{a}{b} - \frac{c}{d} = \frac{a}{b} + \left(-\frac{c}{d}\right) = \frac{ad-bc}{bd} \tag{1.12}$$

Adding the fractions (1.10) can usually be simplified by first finding the "least common denominator" $\text{lcd}(b, d)$ of b and d and then writing the the two fractions with this lcd.

Example 1.18
Find the difference

$$-\frac{5}{14} - \frac{7}{60}$$

SOLUTION:

1. Find the least common denominator (multiple) of 14 and 60:

$$14 = 2^1 \cdot 3^0 \cdot 5^0 \cdot 7^1$$
$$60 = 2^2 \cdot 3^1 \cdot 5^1 \cdot 7^0$$

   hence

$$\text{lcd}(14, 60) = 2^2 \cdot 3^1 \cdot 5^1 \cdot 7^1 = 420$$

2. Write each fraction with the least common denominator of 420 and add the fractions using (1.9):

$$-\frac{5}{14} - \frac{7}{60} = -\frac{5 \cdot 2^1 \cdot 3^1 \cdot 5^1}{14 \cdot 2^1 \cdot 3^1 \cdot 5^1} - \frac{7 \cdot 7^1}{60 \cdot 7^1}$$
$$= -\frac{150}{420} - \frac{49}{420} = -\frac{199}{420}$$

◁

Definition 1.22 (Order Relation)
For rational numbers $a/b, c/d \in \mathbb{Q}$ we define

$$\frac{a}{b} \leq \frac{c}{d} \quad \text{if and only if} \quad ad \leq cb \tag{1.13}$$

Example 1.19
Order the following pairs of rational numbers:
(a) $\frac{1}{2}, \frac{3}{4}$, (b) $-\frac{3}{5}, \frac{4}{7}$, (c) $-\frac{3}{5}, -\frac{6}{7}$

SOLUTION: (a) $\frac{1}{2} \leq \frac{3}{4}$, since $1 \cdot 4 = 4 \leq 6 = 2 \cdot 3$
(b) Recall that $-\frac{3}{5} = \frac{-3}{5}$. Hence $-\frac{3}{5} \leq \frac{4}{7}$, since $(-3) \cdot 7 = -21 \leq 20 = 4 \cdot 5$
(c) Similarly, $-\frac{3}{5} \geq -\frac{6}{7}$, since $-21 \geq -30$. ◁

Theorem 1.1 (Linear Order, Monotony Laws)
Let $a, b, c \in \mathbb{Q}$, then

1. $a \leq a$ *(reflexive)*
2. *If* $a \leq b$ *and* $b \leq a$ *then* $a = b$ *(anti-symmetric)*
3. *If* $a \leq b$ *and* $b \leq c$ *then* $a \leq c$ *(transitive)*
4. $a \leq b$ *or* $b \leq a$ *(dichotomic)*
5. $\leq$ *satisfies the Monotony Laws: For all* $a, b, c \in \mathbb{Q}$,
   (a) *If* $a \leq b$ *then* $a + c \leq b + c$ *for all* $a, b \in \mathbb{Q}$.
   (b) *If* $a \leq b$ *and* $c > 0$ *then* $a \cdot c \leq b \cdot c$

1.1.3 The Linearly Ordered Field of Real Numbers

Recall that the rational numbers are contained in the real numbers, in fact we have the inclusion

$$\mathbb{N} \subseteq \mathbb{Z} \subseteq \mathbb{Q} \subseteq \mathbb{R}$$

The following theorem states the basic algebraic and order theoretic properties of $\mathbb{R}$:

Theorem 1.2 (Algebraic and Order Properties of the Real Numbers)

1. $(\mathbb{R}, +)$ *is a commutative group:*
   (a) $a+(b+c)=(a+b)+c$ *for all* $a, b, c \in \mathbb{R}$, *("+" is associative).*
   (b) $0+a=a+0=a$ *for* $a \in \mathbb{R}$, *(0 is the neutral element with respect to "+").*
   (c) *For each* $a \in \mathbb{R}$ *there exists a (unique) element* $b \in \mathbb{R}$ *such that* $a+b=b+a=0$. *This unique element* b *is called the* additive inverse *of* a *and denoted by* $-a$.
   (d) $a+b=b+a$ *for all* $a, b \in \mathbb{R}$, *(commutative law).*
2. $(\mathbb{R} \setminus \{0\}, \cdot)$ *is a commutative group:*
   (a) $a \cdot (b \cdot c) = (a \cdot b) \cdot c$ *for all* $a, b, c \in \mathbb{R} \setminus \{0\}$, *("·' is associative).*
   (b) $1 \cdot a = a \cdot 1 = a$ *for all* $a \in \mathbb{R} \setminus \{0\}$, *(1 is the neutral element with respect to "·").*
   (c) *For each* $a \in \mathbb{R} \setminus \{0\}$ *there exists a unique element* $b \in \mathbb{R} \setminus \{0\}$ *such that* $a \cdot b = b \cdot a = 1$. *This unique element* b *is called the* multiplicative inverse *of* a *and denoted by* a^{-1} *or* $\frac{1}{a}$.
   (d) $a \cdot b = b \cdot a$ *for* $a, b \in \mathbb{R} \setminus \{0\}$, *("·" is commutative).*
3. Distributive Law: $a \cdot (b+c) = a \cdot b + a \cdot c$ *for all* $a, b, c \in \mathbb{R}$.
4. $(\mathbb{R}, \leq)$ is a linearly ordered set: *Let* $a, b, c \in \mathbb{R}$, *then*
   (a) $a \leq a$ *(reflexive)*
   (b) *If* $a \leq b$ *and* $b \leq a$ *then* $a = b$ *(anti-symmetric)*
   (c) *If* $a \leq b$ *and* $b \leq c$ *then* $a \leq c$ *(transitive)*
   (d) $a \leq b$ *or* $b \leq a$ *(dichotomic)*
   (e) $\leq$ *satisfies the Monotony Laws: For all* $a, b, c \in \mathbb{R}$,
      i. *If* $a \leq b$ *then* $a+c \leq b+c$
      ii. *If* $a \leq b$ *and* $c > 0$ *then* $a \cdot c \leq b \cdot c$

Definition 1.23 (Subtraction, Division, "$<$")

1. For all $a, b, c \in \mathbb{R}$ with $c \neq 0$ we define $a - b := a + (-b)$ and $\frac{a}{c} := a \cdot c^{-1}$
2. For $a, b \in \mathbb{R}$, we define $a < b$ to be equivalent to $a \leq b$ and $a \neq b$.

Example 1.20
Given the binomial formula $(a + b)^2 = a^2 + 2ab + b^2$, find the expansion of $(a - b)^2$.

SOLUTION: Using the definition of subtraction and the given binomial expansion, we first obtain

$$(a - b)^2 = \big(a + (-b)\big)^2 = a^2 + 2a(-b) + (-b)^2$$

and with Corollary 1.1:4d

$$= a^2 - 2ab + b^2$$

◁

Corollary 1.1 (Rules for Manipulating Terms)

1. Cancellation Laws for Addition and Multiplication: *Let $a, b, c \in \mathbb{R}$, then*
   (a) *If $a + c = b + c$ then $a = b$*
   (b) *If $a \cdot c = b \cdot c$ then $a = b$, provided $c \neq 0$,*
2. $a \cdot 0 = 0$ *for all $a \in \mathbb{R}$.*
3. $a \cdot b = 0$ *if and only if $a = 0$ or $b = 0$ for all $a, b \in \mathbb{R}$.*
4. Laws for the Inverse: *Let $a, b \in \mathbb{R}$, then*
   (a) $-a = (-1) \cdot a$
   (b) $-(a + b) = (-a) + (-b) = -a - b$
   (c) $\dfrac{(-a)}{b} = \dfrac{a}{(-b)} = -\dfrac{a}{b}$, *provided $b \neq 0$*
   (d) $(-a) \cdot b = a \cdot (-b) = -(a \cdot b)$
   (e) $(-a) \cdot (-b) = a \cdot b$
5. Fractions of Real Numbers: *Let $a, b, c, d \in \mathbb{R}$ with $b, d \neq 0$, then*
   (a) $\dfrac{a}{b} + \dfrac{c}{d} = \dfrac{a \cdot d + c \cdot b}{b \cdot d}$
   (b) $\dfrac{a}{b} \cdot \dfrac{c}{d} = \dfrac{a \cdot c}{b \cdot d}$

Corollary 1.2 (Rules for Manipulating Inequalities)
Suppose $a, b, c, d \in \mathbb{R}$, then

1. *If $a < b$ then $a + c < b + c$*
2. *If $a \leq b$ and $c \leq d$ then $a + c \leq b + d$*
3. *If $a < b$ and $c \leq d$ then $a + c < b + d$*
4. *If $a < b$ and $0 < c$ then $ac < bc$*
5. *If $0 \leq a \leq b$ and $0 \leq c \leq d$ then $ac \leq bd$*
6. *If $0 \leq a < b$ and $0 < c \leq d$ then $ac < bd$*
7. *If $a \leq b$ and $c < 0$ then $ac \geq bc$*
8. *If $a < b$ and $c < 0$ then $ac > bc$*
9. *If $0 < a$ then $0 < \frac{1}{a}$*
10. *If $0 < a < b$ then $0 < \frac{1}{b} < \frac{1}{a}$*
11. $0 < 1$

Example 1.21
Find the solution set of the inequality $|3x - 1| < 2$.

SOLUTION: $|3x - 1| < 2$ is equivalent to $-2 < 3x - 1 < 2$. In order to isolate x, we first apply Theorem 1.2:4(e)i and add 1 to each component of the double-inequality, which implies $-1 < 3x < 3$. Applying 4(e)ii of the same theorem and multiplying each component of the double-inequality by $\frac{1}{3}$, yields $-\frac{1}{3} < x < 1$. The solution set of the given inequality is therefore $S := (-\frac{1}{3}, 1)$. ◁

1.1.4 The Completeness of the Field of Real Numbers

Definition 1.24 (Upper/Lower Bounds)
Let $S \subseteq \mathbb{R}$ be a nonempty set of real numbers and $b \in \mathbb{R}$.

1. b is said to be an *upper bound* of S if $s \leq b$ for all $s \in S$.
2. b is said to be a *lower bound* of S if $b \leq s$ for all $s \in S$.
3. S is said to be *bounded from above* if it possesses an upper bound.
4. S is said bo be *bounded from below* if it possesses a lower bound.
5. S is said to be *bounded* if it is bounded from above and below.

Definition 1.25 (Maximum, Minimum)
Let $S \subseteq \mathbb{R}$ be a nonempty set of real numbers, and $b \in S$.

1. b is said to be the *maximum* of S if $s \leq b$ for all $s \in S$. The maximum of S is unique and denoted by $\max(S)$.

2. b is said to be the *minimum* of S if $b \leq s$ for all $s \in S$. The minimum of S is unique and denoted by $\min(S)$.

Remark 1.5
Note that the minimum and the maximum of a set are elements of the set. This is different in the case of supremum and infimum of a set as we will see below.

Example 1.22
Find an upper/lower bound and maximum/minimum of the $S := (0, 1]$.

SOLUTION: Clearly, 5 is an upper bound and -1 a lower bound of $(0, 1]$. Moreover, $\max\big((0, 1]\big) = 1$, $(0, 1]$ does not possess a minimum. Why? ◁

Definition 1.26 (Supremum/Infimum, etc.)
Let $S \subseteq \mathbb{R}$ be a nonempty set of real numbers and $b \in \mathbb{R}$.

1. b is called the *least upper bound* or *supremum* of S if

   (a) b is an upper bound of S and

   (b) if $b' \in \mathbb{R}$ is any upper bound of S then $b \leq b'$.

   The least upper bound or supremum of S is unique and is denoted by $\text{lub}(S)$ or $\sup(S)$.

2. b is called the *greatest lower bound* or *infimum* of S if

   (a) b is a lower bound of S and

   (b) if $b' \in \mathbb{R}$ is any lower bound of S then $b \geq b'$.

   The greatest lower bound or infimum of S is unique and denoted by $\text{glb}(S)$ or $\inf(S)$.

Remark 1.6
Note that if $\sup(S) \in S$, then $\sup(S) = \max(S)$. Similarly, if $\inf(S) \in S$, then $\inf(S) = \min(S)$.

Example 1.23
Find max, min, sup, inf of $S_1 := \{1/n \mid n \in \mathbb{N}\}$

SOLUTION: Clearly, $S_1 = \{1, 1/2, 1/3, 1/4, \ldots\}$ is bounded from above and below, for instance by 2 and -1. Note that $\inf(S_1) = 0$ does not belong to the set S_1, hence $\min(S_1)$ does not exist. On the other hand $\sup(S_1) = \max(S_1)$. ◁

Example 1.24
Determine whether the set $S := \{(x+1)/x \mid x \in (0,\infty)\}$ is bounded and determine $\sup(S)$, $\inf(S)$ if they exist.

SOLUTION: Clearly,

$$\lim_{x\to 0+} \frac{x+1}{x} = \lim_{x\to 0+} \left(1 + \frac{1}{x}\right) = \infty,$$

so that S is not bounded from above. Hence $\sup(S)$ does not exist. On the other hand, S is bounded from below, for instance, by 1, which is the greatest lower bound, i.e. $\inf(S) = 1$. Note that $1 \notin S$. Therefore, $\min(S)$ does not exist! $\triangleleft$

Theorem 1.3 (Completeness of $\mathbb{R}$, Least Upper Bound Axiom)
If $S \subseteq \mathbb{R}$ is nonempty and bounded above then S possesses a least upper bound in $\mathbb{R}$.

Remark 1.7
Both $\mathbb{Q}$ and $\mathbb{R}$ are linearly ordered fields, i.e. they satisfy properties 1, 2, 3 and 4 of Theorem 1.2. However, while $\mathbb{Q}$ possesses only countably infinitely many elements, $\mathbb{R}$ is uncountably infinite, since $\mathbb{R}$ in addition satisfies the completeness property stated in Theorem 1.3.

Example 1.25
Show that $\sqrt{2} \notin \mathbb{Q}$, i.e. that $\sqrt{2}$ is *not* a rational number.

SOLUTION: Suppose $\sqrt{2} \in \mathbb{Q}$. Then there exist $m, n \in \mathbb{Z}$ such that

$$\sqrt{2} = \frac{m}{n}$$

where m, n can be chosen to be relatively prime, i.e. their greatest common divisor is 1. We first square this equation

$$2 = \frac{m^2}{n^2}$$

and then multiply it by n^2 yielding

$$2n^2 = m^2$$

Clearly, m^2 is divisible by 2, and therefore (by Euclid's Lemma) also m. Hence, $m = 2\tilde{m}$. Replacing m by $2\tilde{m}$, we obtain

$$2n^2 = 4\tilde{m}^2$$

Dividing the new equation by 2

$$n^2 = 2\tilde{m}^2$$

we see that also n^2 and therefore n is divisible by 2. I.e. both m and n are divisible by 2, in contradiction to m and n are relatively prime. Hence, our assumption $\sqrt{2} \in \mathbb{Q}$ is false, which implies that $\sqrt{2} \notin \mathbb{Q}$. $\triangleleft$

Theorem 1.4 (Existence of n-th Roots in $\mathbb{R}$)
For every real number $a \in \mathbb{R}$, $a > 0$ and every natural number $n \in \mathbb{N}$ there exists a unique real number $x \in \mathbb{R}$, $x > 0$ such that

$$x^n = a$$

This number x is denoted by

$$\sqrt[n]{a} \quad \text{or} \quad a^{1/n}$$

i.e. $x = \sqrt[n]{a} = a^{1/n}$.

PROOF: (1) *Uniqueness:* Since $0 < x_1 < x_2$ implies that $0 < x_1^n < x_2^n$, there can at most be one such x.

(2) *Existence:* Consider the set

$$X := \{t \in \mathbb{R} \mid t > 0,\ t^n < a\}$$

(a) $E \neq \emptyset$: consider $t := a/(1+a)$. Clearly, $0 < t < 1$, hence $t^n < t < a$ and thus $t \in E$, i.e. $E \neq \emptyset$.

(b) E is bounded from above: E is bounded by $1 + a$. If $t > 1 + a > 1$ then $t^n > t > 1 + a > a$, hence $t \notin E$.

Hence, by Theorem 1.3, $\sup(E)$ exists. Let

$$x := \sup(E)$$

To verify that $x^n = a$, show that both $x^n < a$ and $x^n > a$ lead to contradictions. We'll omit this technical part of the proof. $\square$

1.1.5 Powers and Radicals

Definition 1.27 (Powers with Integer Exponents)
For $a \in \mathbb{R}$ and $n \in \mathbb{N}^0$ we define the *power* a^n recursively by

$$a^0 := 1 \quad \text{and} \quad a^{n+1} := a^n \cdot a,$$

i.e. if $n > 0$, we have

$$a^n = \underbrace{a \cdot a \cdots a}_{n \text{ times}}.$$

If $a \in \mathbb{R} \setminus \{0\}$ and $n \in \mathbb{N}$, then we define

$$a^{-n} := (a^{-1})^n = (a^n)^{-1}.$$

The number a is called the *base,* while the number n is called the *exponent* of the power.

Example 1.26
Find 5^3 and 5^{-3}.

SOLUTION:

$$\begin{aligned} 5^3 &= 5 \cdot 5 \cdot 5 = 125 \\ 5^{-3} &= \frac{1}{5} \cdot \frac{1}{5} \cdot \frac{1}{5} \\ &= \frac{1}{5 \cdot 5 \cdot 5} = \frac{1}{125} \end{aligned}$$

◁

Theorem 1.5 (Real Roots)
For any $n \in \mathbb{N}$ and any positive $a \in \mathbb{R}$ there exists precisely one positive $r \in \mathbb{R}$ such that

$$r^n = a.$$

This unique number r is called the principal n-th root *of a and denoted by*

$$\sqrt[n]{a} := a^{\frac{1}{n}} := r.$$

If $n \in \mathbb{N}$ is odd, we define $\sqrt[n]{-a} := -\sqrt[n]{a}$. The symbol $\sqrt[n]{a}$ is called a radical symbol, *the number n the* index of the radical *and the number a the* radicant.

Example 1.27
Simplify and rationalize the expression $\left(5^{1/5}\right)^{-2}$.

SOLUTION:

$$\begin{aligned}\left(5^{1/5}\right)^{-2} &= \frac{1}{\left(5^{1/5}\right)^2} = \frac{1}{5^{2/5}}\\ &= \frac{1}{5^{2/5}} \cdot \frac{5^{3/5}}{5^{3/5}} = \frac{1 \cdot 5^{3/5}}{5} = \tfrac{1}{5}5^{3/5}\end{aligned}$$

◁

Theorem 1.6 (Complex Roots)
For any $n \in \mathbb{N}$ and any complex number $z = r(\cos\theta + i\sin\theta) \in \mathbb{C}$, $z \neq 0$, there exist precisely n numbers ω_k $(k = 0, 1, \ldots, n-1)$ such that

$$\omega_k^n = z \quad k = 0, 1, \ldots, n-1,$$

namely,

$$\begin{aligned}\omega_k &= r^{\frac{1}{n}}\left(\cos\left(\tfrac{\theta}{n} + \tfrac{2\pi k}{n}\right) + i\sin\left(\tfrac{\theta}{n} + \tfrac{2\pi k}{n}\right)\right)\\ &= r^{\frac{1}{n}} e^{i\frac{\theta+2\pi k}{n}}\end{aligned}$$

*for $k = 0, 1, \ldots, n-1$. The numbers ω_k are called n-*th roots *of z. If $z = 1$, they are called n-*th roots of unity.

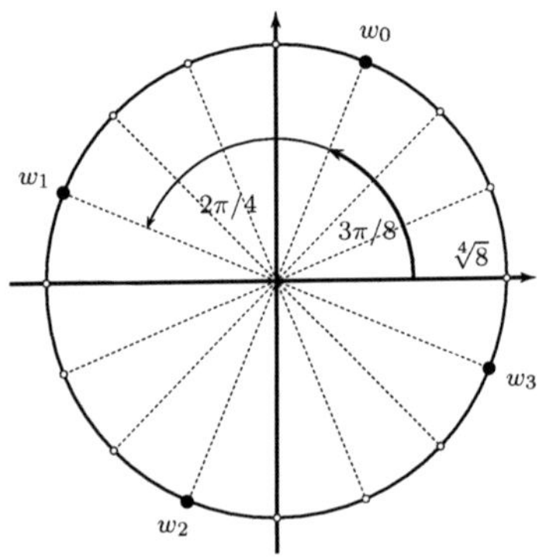

Figure 1.17: *The four roots of $-8i$*

Example 1.28
Compute the 4-th roots of $-8i$.

SOLUTION: First note that the argument of $-8i$ is $3\pi/2$ and $|-8i| = 8$. Thus

$$-8i = 8\left(\cos\tfrac{3\pi}{2} + i\sin\tfrac{3\pi}{2}\right).$$

Hence, by Theorem 1.6, the four fourth roots of $-8i$ are given by

$$w_k = 8^{1/4}\left[\cos\tfrac{1}{4}\left(\tfrac{3\pi}{2} + 2k\pi\right) + i\sin\tfrac{1}{4}\left(\tfrac{3\pi}{2} + 2k\pi\right)\right]$$

for $k = 0, 1, 2, 3$, or explicitly

$$\begin{aligned}w_0 &= 8^{1/4}\left[\cos\tfrac{3\pi}{8} + i\sin\tfrac{3\pi}{8}\right]\\ w_1 &= 8^{1/4}\left[\cos\tfrac{7\pi}{8} + i\sin\tfrac{7\pi}{8}\right]\\ w_2 &= 8^{1/4}\left[\cos\tfrac{11\pi}{8} + i\sin\tfrac{11\pi}{8}\right]\\ w_3 &= 8^{1/4}\left[\cos\tfrac{15\pi}{8} + i\sin\tfrac{15\pi}{8}\right]\end{aligned}$$

◁

Definition 1.28 (Powers with Rational Exponents)
For any positive real number $a \in \mathbb{R}$ and any rational number $\frac{n}{m} \in \mathbb{Q}$, we define

$$a^{\frac{n}{m}} := \sqrt[m]{a^n} = \left(\sqrt[m]{a}\right)^n.$$

Example 1.29
Find $8^{-\frac{2}{3}}$!

SOLUTION: By definition,

$$8^{-\frac{2}{3}} = 8^{\frac{-2}{3}} = \sqrt[3]{8^{-2}} = \sqrt[3]{\frac{1}{8^2}} = \sqrt[3]{\frac{1}{64}} = \frac{1}{4}.$$

◁

Definition 1.29 (Powers with Real Exponents)
For any positive real number $a \in \mathbb{R}$ and any real number $r \in \mathbb{R}$, we define

$$a^r := e^{r \ln a}.$$

where $e \in \mathbb{R}$ is the Euler constant. Recall that

$$e^x = \sum_{n=0}^{\infty} \frac{x^n}{n!} = 1 + x + \frac{x^2}{2!} + \frac{x^3}{3!} + \cdots$$

for all $x \in \mathbb{R}$.

Example 1.30
Evaluate the power $\sqrt{3}^{\sqrt{2}}$!

SOLUTION: Again, by definition, we have

$$\sqrt{3}^{\sqrt{2}} = e^{\sqrt{2}\ln(\sqrt{3})}$$

.

◁

Theorem 1.7 (Laws of Exponent)
Let $a, b \in \mathbb{R}$ and $r, s \in \mathbb{R}$, then

1. $a^r \cdot a^s = a^{r+s}$
2. $\dfrac{a^r}{a^s} = a^{r-s} = \dfrac{1}{a^{s-r}}$
3. $a^r \cdot b^r = (a \cdot b)^r$
4. $\dfrac{a^r}{b^r} = \left(\dfrac{a}{b}\right)^r, \quad b \neq 0$
5. $(a^r)^s = a^{rs}$
6. *For all $n \in \mathbb{Z}$ we have*

$$(-1)^n = \begin{cases} 1, & \text{if } n \text{ is even} \\ -1, & \text{if } n \text{ is odd} \end{cases}$$

Corollary 1.3 (Laws of Radical)
Let $a, b \in \mathbb{R}$, $a, b > 0$ and $n, m \in \mathbb{N}$, then

1. $\sqrt[n]{1} = 1, \quad \sqrt[n]{0} = 0$
2. $\sqrt[m]{a^n} = \left(\sqrt[m]{a}\right)^n = a^{\frac{n}{m}}$
3. $\sqrt[n]{a} \cdot \sqrt[n]{b} = \sqrt[n]{a \cdot b}$
4. $\dfrac{\sqrt[n]{a}}{\sqrt[n]{b}} = \sqrt[n]{\dfrac{a}{b}}$
5. $\sqrt[n]{\sqrt[m]{a}} = \sqrt[m]{\sqrt[n]{a}} = \sqrt[nm]{a}$
6. $\left(\sqrt[n]{a}\right)^n = a$
7. $\sqrt[n]{a^n} = \begin{cases} |a| & \text{if } n \text{ is even} \\ a & \text{if } n \text{ is odd} \end{cases}$

1.1.6 Absolute Value

Definition 1.30 (Absolute Value)
For any $x \in \mathbb{R}$ the number

$$|x| := \begin{cases} x, & \text{if } x \geq 0, \\ -x, & \text{if } x < 0 \end{cases}$$

is called the *absolute value* of x.

Theorem 1.8 (Properties of $|\cdot|$)
For all $x, y \in \mathbb{R}$,

1. $|x| \geq 0$ *and* $\big(|x| = 0 \Leftrightarrow x = 0\big)$
2. $|xy| = |x||y|$
3. $|x + y| \leq |x| + |y|$ *(triangle inequality)*

Lemma 1.6
Suppose $x, a, \varepsilon \in \mathbb{R}$ and $\varepsilon > 0$, then

1. $|x| < \varepsilon \Leftrightarrow -\varepsilon < x < \varepsilon$
2. $|x - a| < \varepsilon \Leftrightarrow a - \varepsilon < x < a + \varepsilon$
3. *The two previous statements remain true if "$<$" is replaced by "$\leq$".*

1.1.7 Factorial and Binomial Coefficients

Definition 1.31
For all $n \in \mathbb{N}^0$, we define the number $n!$ (read: n factorial) recursively by

$$0! := 1 \quad \text{and} \quad (n+1)! := n!(n+1),$$

i.e.: $n! = 1 \cdot 2 \cdot 3 \cdots (n-1) \cdot n$

Example 1.31
Compute 3! and 4!.

SOLUTION: $3! = 1 \cdot 2 \cdot 3 = 6$, $4! = 3! \cdot 4 = 6 \cdot 4 = 24$. $\triangleleft$

Definition 1.32 (Binomial Coefficients)
For all $a \in \mathbb{R}$ and $k \in \mathbb{N}^0$ we define the *Binomial coefficient* $\binom{a}{k}$ recursively by

$$\binom{a}{0} := 1 \quad \text{and} \quad \binom{a}{k+1} := \binom{a}{k} \frac{a-k}{k+1},$$

i.e. for $k > 0$

$$\binom{a}{k} = \frac{a}{1} \frac{(a-1)}{2} \frac{(a-2)}{3} \cdots \frac{(a-k+1)}{k}$$

Example 1.32
Compute the binomial coefficients $\binom{2}{k}$ and $\binom{3}{k}$ for $k = 0, 1, 2, 3$.

SOLUTION:

$$\binom{2}{0} = 1$$
$$\binom{2}{1} = \frac{2}{1} = 2$$
$$\binom{2}{2} = \frac{2}{1} \frac{(2-1)}{2} = 1$$

$$\binom{3}{0} = 1$$
$$\binom{3}{1} = \frac{3}{1} = 3$$
$$\binom{3}{2} = \frac{3}{1} \frac{(3-1)}{2} = 3$$
$$\binom{3}{3} = \frac{3}{1} \frac{(3-1)}{2} \frac{(3-2)}{3} = 1$$

◁

Lemma 1.7
Let $k, n \in \mathbb{N}^0$, then

1. $\binom{n}{k} = \dfrac{n!}{k!(n-k)!}$, *for* $k \leq n$.
2. $\binom{n}{k} = \binom{n}{n-k}$, *for* $k \leq n$.
3. $\binom{n}{k} + \binom{n}{k+1} = \binom{n+1}{k+1}$, $k \leq n-1$.
4. $\binom{n}{k} = 0$, *for* $k > n$.

Remark 1.8 (Pascal's Triangle)

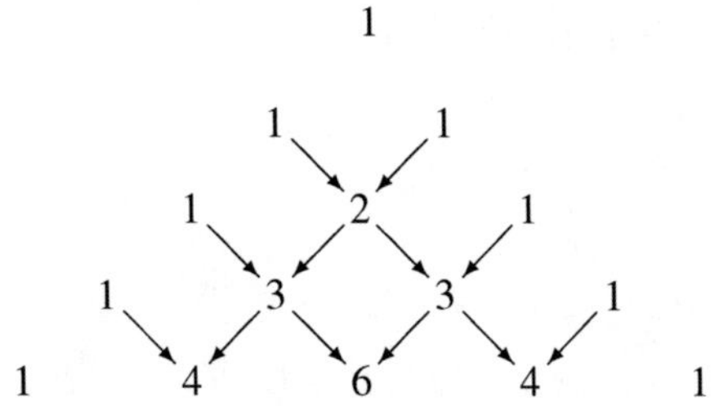

The trianglular arrangment of the binomial coefficients $\binom{n}{k}$ suggested by the recursion formula 3 allows a systematic computation of the binomial coefficients starting with $\binom{0}{0} = 1$. The coefficients $\binom{n}{0} = \binom{n}{n}$ $(n \in \mathbb{N}^0)$ forming the sides of the triangle are all 1, while the coefficients in the interior are the sum of the two binomial coefficients directly above as indicated by the arrows in the diagram above.

1.1.8 Binomial Formulas, Sums/Differences of Powers

Theorem 1.9 (Binomial Formula)
For all $a, b \in \mathbb{R}$ and $n \in \mathbb{N}$, we have

$$(a+b)^n = \sum_{k=0}^{n} \binom{n}{k} a^{n-k} b^k$$

in particular

$$\begin{aligned}(a+b)^2 &= a^2 + 2ab + b^2 \\ (a+b)^3 &= a^3 + 3a^2b + 3ab^2 + b^3 \\ (a+b)^4 &= a^4 + 4a^3b + 6a^2b^2 + 4ab^3 + b^4\end{aligned}$$

Example 1.33
Expand the binomials $(a-b)^2$ and $(a-b)^3$

SOLUTION: We apply Definition 1.23:1

$$(a-b)^2 = \big(a + (-b)\big)^2 = a^2 + 2a(-b) + (-b)^2 = a^2 - 2ab + b^2$$

Similarly

$$\begin{aligned}(a-b)^3 = \big(a + (-b)\big)^3 &= a^3 + 3a^2(-b) + 3a(-b)^2 + (-b)^3 \\ &= a^3 - 3a^2b + 3ab^2 - b^3\end{aligned}$$

◁

Example 1.34
Expand the binomial expression $\left(2x^3y - z^2\right)^2$

SOLUTION:

$$\begin{aligned}\left(2x^3y - z^2\right)^2 &= \left(2x^3y + (-z^2)\right)^2 = (2x^3y)^2 + 2(2x^3y)(-z^2) + (-z^2)^2 \\ &= 2^2(x^3)^2y^2 - 4x^3yz^2 + (z^2)^2 = 4x^6y^2 - 4x^3yz^2 + z^4\end{aligned}$$

◁

Lemma 1.8 (Completing the Square)
If , $x \in \mathbb{R}$, *then*

$$x^2 + 2ax = (x + a)^2 - a^2$$

Remark 1.9
Note that a is half the coefficient of the linear term on the left side!

Example 1.35
Solve the quadratic equation $x^2 + 4x - 7 = 0$ by completing the square.

SOLUTION: $0 = x^2 + 4x - 7 = [x^2 + 4x] - 7 = [(x + 2)^2 - 2^2] - 7 = (x + 2)^2 - 11$ which is equivalent to $(x + 2)^2 = 11$. Taking the square root on both sides, we obtain $|x + 2| = \sqrt{11}$ and thus $x + 2 = \pm\sqrt{11}$. Hence, $x = -2 \pm \sqrt{11}$. ◁

Theorem 1.10 (Differences of Powers)
For all $a, b \in \mathbb{R}$ *and* $n \in \mathbb{N}$, *we have*

$$a^n - b^n = (a - b)\sum_{k=0}^{n-1} a^{n-1-k}b^k$$

in particular

$$\begin{aligned}a^2 - b^2 &= (a - b)(a + b) \\ a^3 - b^3 &= (a - b)(a^2 + ab + b^2) \\ a^4 - b^4 &= (a - b)(a^3 + a^2b + ab^2 + b^3)\end{aligned}$$

Theorem 1.11 (Sums of Powers)
For all $a, b \in \mathbb{R}$ and $n \in \mathbb{N}$, we have

$$a^{2n+1} + b^{2n+1} = (a+b)\sum_{k=0}^{2n}(-1)^k a^{2n-k}b^k$$

in particular

$$\begin{aligned} a^3 + b^3 &= (a+b)(a^2 - ab + b^2) \\ a^5 + b^5 &= (a+b)(a^4 - a^3 b \\ &\qquad + a^2b^2 - ab^3 + b^4) \\ a^7 + b^7 &= (a+b)(a^6 - a^5 b + a^4 b^2 \\ &\qquad - a^3b^3 + a^2b^4 - ab^5 + b^6) \end{aligned}$$

1.1.9 Exponential and Logarithmic Functions

Definition 1.33 (Exponential Function with Base a)
For $a \in \mathbb{R}$, $a > 0$ and $a \neq 1$, the association $x \mapsto a^x$ defines a functions $\exp_a : \mathbb{R} \to \mathbb{R}$ called the *base-a exponential function.* In the special and important case where $a = e$, the Euler constant, *exp* $:= \exp_e$ is called the *natural exponential function.* (cf. Defintion 1.29),

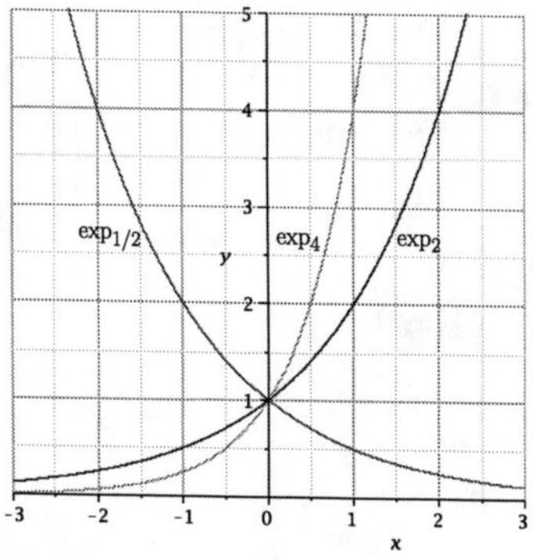

Figure 1.18: *Graphs of* $\exp_{1/2}$, $\exp_2$ *and* $\exp_4$

Note that $\exp_a$ is increasing for $a > 1$ and decreasing for $0 < a < 1$. Also note that all exponential functions have y-intercept $y = 0$.

Example 1.36
Sketch the graphs of the exponential functions

$$\begin{aligned} \exp_{1/2}(x) &:= \left(\tfrac{1}{2}\right)^x \\ \exp_2(x) &:= 2^x \\ \exp_4(x) &:= 4^x \end{aligned}$$

SOLUTION: ◁

Theorem 1.12 (Properties of $\exp_a$)
Let $a \in \mathbb{R}$, $a > 0$, $a \neq 1$. For all $x_1, x_2 \in \mathbb{R}$:

1. $\exp_a(0) = 1$
2. $\exp_a(x_1 + x_2) = \exp_a(x_1)\exp_a(x_2)$
3. $\exp_a(x_1) = \exp_a(x_2) \Rightarrow x_1 = x_2$
   i.e. $\exp_a$ is injective (one-to-one) and therefore invertible. Its inverse is the logarithmic function $\log_a$.
4. *If $0 < a < 1$ then $\exp_a$ is decreasing*
   If $a > 1$ then $\exp_a$ is increasing.
5. *$\exp_a$ has domain $\mathbb{R}$*
6. *$\exp_a$ has range $(0, \infty)$*

Remark 1.10
Because of Theorem 1.12:1,2, $\exp_a(x)$ is usually written as a^x. Therefore, these equations can be written in the more familar form:

1. $a^0 = 1$
2. $a^{x_1 + x_2} = a^{x_1}a^{x_2}$
3. $a^{x_1} = a^{x_2} \Rightarrow x_1 = x_2$

Definition 1.34 (Logarithmic Functions)
Let $a \in \mathbb{R}$, $a > 0$, $a \neq 1$. The *logarithmic function* $\log_a : (0, \infty) \to \mathbb{R}$ with base a is the inverse of the base-a exponential function $\exp_a : \mathbb{R} \to \mathbb{R}$ and is implicitly defined by

$$\log_a(x) = y \Leftrightarrow \exp_a(y) = x$$
$$\Leftrightarrow a^y = x$$

for all $x \in (0, \infty)$ and $y \in \mathbb{R}$. The logarithm with base e, $\log_e$, is called the *natural logarithm* and is usually denoted by ln

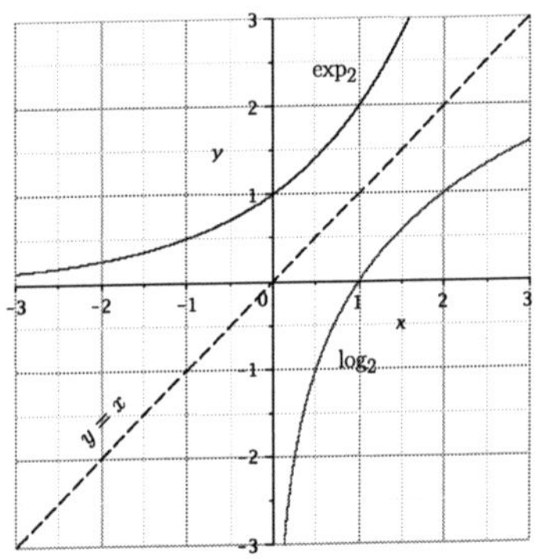

Figure 1.19: *Graphs of* $\log_2$ *and* $\exp_2$

Example 1.37
Sketch the graphs of $\log_2$ and $\exp_2$.

SOLUTION: ◁

Theorem 1.13 (Properties of $\log_a$)

1. $\log_a(1) = 0$
2. $\log_a(x_1 x_2) = \log_a(x_1) + \log_a(x_2)$
   $\left[\log_a\left(\frac{x_1}{x_2}\right) = \log_a(x_1) - \log_a(x_2)\right]$
3. $\log_a(x^r) = r\log_a(x)$
   $\left[\log_a(a^r) = r\right]$
4. $\log_a(x_1) = \log_a(x_2) \Rightarrow x_1 = x_2$
   i.e. $\log_a$ is injective (one-to-one) and therefore invertible. Its inverse is the exponential function $\exp_a$. In particular, we have the cancellation equations:

$$\exp_a\Big(\log_a(x)\Big) = x \quad \forall x \in (0, \infty)$$
$$\log_a\Big(\exp_a(x)\Big) = x \quad \forall x \in \mathbb{R}$$

   or in simplified notation:

$$a^{\log_a x} = x \quad \forall x \in (0, \infty)$$
$$\log_a a^x = x \quad \forall x \in \mathbb{R}$$

5. *$\log_a$ has domain $(0, \infty)$*
6. *$\log_a$ has range $\mathbb{R}$*
7. *Base Change Formula: Suppose $a, b \in \mathbb{R}$, $a, b > 0$, $a, b \neq 1$. then*

$$\log_a(x) = \frac{\log_b(x)}{\log_b(a)}$$

   for all $x \in (0, \infty)$.

Example 1.38
Evaluate $\log_6 5$ correctly to nine decimal places.

SOLUTION: Using the base change formula, we obtain

$$\log_6 5 = \frac{\ln 5}{\ln 6} \approx 0.898\,244\,401\,7$$

◁

1.1.10 Polynomials

Definition 1.35 (Polynomial)
An expression of the form

$$p(x) = a_0 + a_1x + a_2x^2 + \cdots + a_nx^n$$
$$= \sum_{k=0}^{n} a_kx^k,$$

is called a *polynomial.* The numbers a_k are called the *coefficients* of p, a_0 the absolute term, a_n the *leading coefficient* and n the *degree* of p, denoted by $\deg(p)$, provided $a_n \neq 0$. p is called a *monic polynomial* if the leading coefficient $a_n = 1$. The symbol x is called an *indeterminant.* Polynomials of degree $0, 1, 2, 3$ are called *constant, linear, quadratic* and *cubic* polynomials, respectively.
Let p be a polynomial. To emphasize that all the coefficents of p are integers, rational numbers, real numbers or complex numbers, we say that p is a *polynomial over* $\mathbb{Z}$, $\mathbb{Q}$, $\mathbb{R}$ or $\mathbb{C}$, respectively.

Example 1.39
Classify the following polynomials and sketch their graph:

$$p_1(x) = x - 3$$
$$p_2(x) = x^2 + (3 - \sqrt{2})x - 3\sqrt{2}$$
$$p_3(x) = \tfrac{1}{2}x^5 - \tfrac{1}{2}x^4 + \tfrac{1}{2}x^3 + 6x^2 - \tfrac{171}{20}x - \tfrac{27}{2}$$

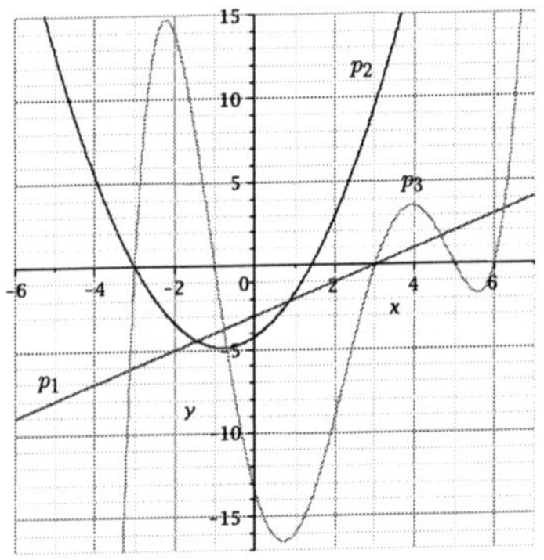

Figure 1.20:

SOLUTION: p_1 is a monic linear polynomial (also called a linear function), p_2 is a quadratic polynomial, and $p_3(x)$ is a polynomial of degree 5. p_1 is a polynomial over $\mathbb{Z}$, p_2 is a monic polynomial over $\mathbb{R}$, p_3 is a polynomial over the field of rational numbers $\mathbb{Q}$. ◁

Linear Functions

Theorem 1.14 (Line Equations)

1. *Slope of a line:* If (x_1, y_1) and (x_2, y_2) are two points of a line ℓ and if $x_1 \neq x_2$, then the slope $m = m(\ell)$ of ℓ is defined by

$$m = m(\ell) = \frac{\Delta y}{\Delta x} = \frac{y_2 - y_1}{x_2 - x_1}$$

2. *Slope-Intercept Form:* If the line ℓ has slope m and y-intercept b, then

$$y = mx + b$$

3. *Point-Intercept Form:* If the line ℓ has slope m and passes through the point (x_1, y_1) then

$$y - y_1 = m(x - x_1)$$

If ℓ is a vertical line, then $y = x_1$.

4. *Intercept Form:* If the line ℓ has x-intercept a and y-intercept b then

$$\frac{x}{a} + \frac{y}{b} = 1$$

5. *Parallel Lines:* If ℓ_1 and ℓ_2 are two lines then

$$\ell_1 \parallel \ell_2 \Leftrightarrow m(\ell_1) = m(\ell_2)$$

6. *Perpendicular Lines:* If ℓ_1 and ℓ_2 are two lines then

$$\ell_1 \perp \ell_2 \Leftrightarrow m(\ell_1) \cdot m(\ell_2) = -1$$

Quadratic Functions

Definition 1.36 (Standard Form of the Quadratic Function)
The quadratic function

$$p(x) = a(x - x_1)^2 + y_1$$

is said to be in *standard form*, $a \neq 0$. Its *line of symmetry* is the vertical line $y = x_1$, its *vertex* is the point (x_1, y_1).

Example 1.40 (Rewriting in Standard Form)
Transform the quadratic function $p(x) = ax^2 + bx + c$, $a \neq 0$, into standard form.

SOLUTION: Using "completing the square" we obtain:

$$\begin{aligned} p(x) &= ax^2 + bx + c \\ &= a\left[x^2 + \tfrac{b}{a}x\right] + c \\ &= a\left[\left(x + \tfrac{b}{2a}\right)^2 - \tfrac{b^2}{4a^2}\right] + c \\ &= a\left(x + \tfrac{b}{2a}\right)^2 + c - \tfrac{b^2}{4a} \\ &= a\Big(x - \underbrace{\left(-\tfrac{b}{2a}\right)}_{x_1}\Big)^2 + \underbrace{\tfrac{4ac-b^2}{4a}}_{y_1} \end{aligned}$$

◁

Example 1.41
Solve the quadratic equation $x^2 + 4x - 7 = 0$.

SOLUTION: We first transform the quadratic function into standard form using "completing the square":
$0 = x^2 + 4x - 7 = [x^2 + 4x] - 7 = [(x+2)^2 - 2^2] - 7 = (x+2)^2 - 11$ which is equivalent to $(x+2)^2 = 11$. Taking the square root on both sides, we obtain $|x+2| = \sqrt{11}$ and thus $x + 2 = \pm\sqrt{11}$. Hence, $x = -2 \pm \sqrt{11}$. ◁

Theorem 1.15 (Quadratic Formula)
Suppose that $p(x) = ax^2 + bx + c$ is a quadratic polynomial. Then its zeroes are

$$x = \frac{-b \pm \sqrt{b^2 - 4ac}}{2a}.$$

Example 1.42
Rewrite the quadratic function $p(x) = x^2 - 6x + 7$ in standard form, find its line of symmetry, vertex and zeroes.

SOLUTION: Completing the square, we have

$$\begin{aligned} p(x) &= x^2 - 6x + 7 \\ &= (x-3)^2 - (-3)^2 + 7 \\ &= (x-3)^2 - 2 \end{aligned}$$

Hence, $x = 3$ is the line of symmetry of p and $(3, -2)$ its vertex. To find its zeroes (x-intercepts), we solve

$$0 = p(x) = (x-3)^2 - 2$$

which is equivalent to

$$(x-3)^2 = 2$$

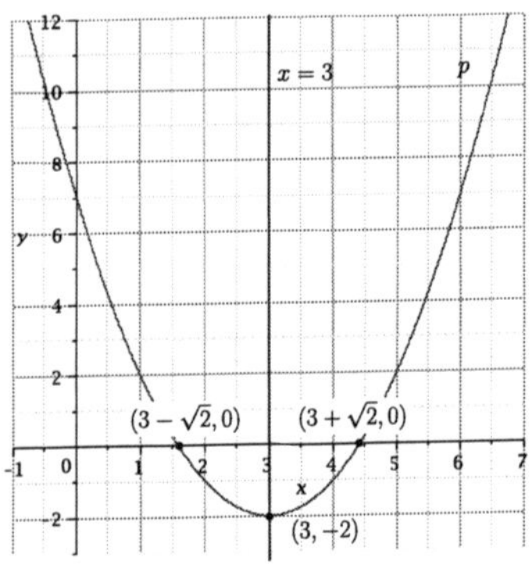

Figure 1.21:

and from which by taking the square root follows

$$x = 3 \pm \sqrt{2}$$

◁

Factoring Polynomials, Zeros

Lemma 1.9 (Factoring Quadratic Polynomials)

1. $x^2 + 2ax + a^2 = (x + a)^2$
2. $x^2 + (a + b)x + ab = (x + a)(x + b)$
3. $a_1 a_2 x^2 + (a_1 b_2 + a_2 b_1)x + b_1 b_2 = (a_1 x + b_1)(a_2 x + b_2)$

Lemma 1.10 (Differences of Powers)
For all $x, a \in \mathbb{R}$ and $n \in \mathbb{N}$, we have

$$x^n - a^n = (x - a) \sum_{k=0}^{n-1} x^{n-1-k} a^k$$

in particular

$$\begin{aligned} x^2 - a^2 &= (x - a)(x + a) \\ x^3 - a^3 &= (x - a)(x^2 + xa + a^2) \\ x^4 - a^4 &= (x - a)(x^3 + x^2 a + xa^2 + a^3) \end{aligned}$$

Lemma 1.11 (Sums of Powers)
For all $x, a \in \mathbb{R}$ and $n \in \mathbb{N}$, we have

$$x^{2n+1} + a^{2n+1} = (x + a) \sum_{k=0}^{2n} (-1)^k x^{2n-k} a^k$$

in particular

$$\begin{aligned} x^3 + a^3 &= (x + a)(x^2 - xa + a^2) \\ x^5 + a^5 &= (x + a)(x^4 - x^3 a + x^2 a^2 - xa^3 + a^4) \\ x^7 + a^7 &= (x + a)(x^6 - x^5 a + x^4 a^2 - x^3 a^3 + x^2 a^4 - xa^5 + a^6) \end{aligned}$$

Theorem 1.16 (Division Algorithm)
If p and d are polynomials with $d \neq 0$ then there exist unique polynomials q and r such that

$$p = dq + r$$

where either $r = 0$ or $\deg(r) < \deg(d)$.

The polynomial p is called dividend, *d the* divisor, *q the* quotient *and r the* remainder.
If $r = 0$, we say that the polynomial d divides *the polynomial p and we write $d \mid p$.*

Example 1.43 (Long Division)
Divide the polynomial $p(x) = 2x^4 + 4x^3 - 5x^2 + 3x - 2$ by the polynomial $d(x) = x^2 + 2x - 3$ using "long division".

SOLUTION:

$$\begin{array}{rl}
 & \phantom{2x^4 + 4x^3 - {}} 2x^2 \phantom{{}+3x} + 1 \\
x^2 + 2x - 3 \,\big) & \overline{2x^4 + 4x^3 - 5x^2 + 3x - 2} \\
 & \underline{2x^4 + 4x^3 - 6x^2} \\
 & \phantom{2x^4 + 4x^3 - 5{}} x^2 + 3x - 2 \\
 & \phantom{2x^4 + 4x^3 - 5{}} \underline{x^2 + 2x - 3} \\
 & \phantom{2x^4 + 4x^3 - 5x^2 + 3{}} x + 1
\end{array}$$

We obtain

$$\begin{aligned}
&\overbrace{2x^4 + 4x^3 - 5x^2 + 3x - 2}^{p(x)} = \\
= \quad &\underbrace{(x^2 + 2x - 3)}_{d(x)}\underbrace{(2x^2 + 1)}_{q(x)} + \underbrace{x + 1}_{r(x)}
\end{aligned}$$

◁

Example 1.44 (Long Division)
Divide the polynomial $p(x) = 6x^3 - 19x^2 + 16x - 7$ by the polynomial $d(x) = x - 2$ using long division.

SOLUTION:

$$\begin{array}{rl}
 & 6x^2 - 7x + 2 \\
x - 2 \,\big) & \overline{6x^3 - 19x^2 + 16x - 7} \\
 & \underline{6x^3 - 12x^2} \\
 & -7x^2 + 16x \\
 & \underline{-7x^2 + 14x} \\
 & \phantom{6x^3 - 19x^2 + {}} 2x - 7 \\
 & \phantom{6x^3 - 19x^2 + {}} \underline{2x - 4} \\
 & - 3
\end{array}$$

Thus

$$\overbrace{6x^3 - 19x^2 + 16x - 7}^{p(x)} = \\ = \underbrace{(x-2)}_{d(x)}\underbrace{(6x^2 - 7x + 2)}_{q(x)} + \underbrace{(-3)}_{r(x)}$$

Since $\deg(x - 2) = 1$ and $\deg r < 1$, the remainder has to be a constant polynomial. Moreover, since $d(2) = 0$, the value of p at 2 is $p(2) = -3$. $\lhd$

Definition 1.37 (Zeros or Roots of a Polynomial)
If p is a polynomial over C, then $\alpha \in \mathbb{C}$ is called a *zero* or *root* of p if $p(\alpha) = 0$.

Corollary 1.4 (The Remainder Theorem)
If p is a polynomial over C and $a \in \mathbb{C}$ then $p(a)$ is the remainder in the division by $x - a$.

PROOF: By the Division Algorithm (Theorem 1.16), there exist unique polynomial p and r with $r = 0$ or $\deg(r) < \deg(d) = 1$ such that

$$p(x) = (x - a)q(x) + r$$

Note that a polynomial of degree less than 1 is a constant! Hence

$$p(a) = (a - a)q(a) + r = 0 + r = r$$

as claimed. $\square$

Corollary 1.5 (The Factor Theorem)
Let p be a polynomial over C and $\alpha \in \mathbb{C}$. Then

$$p(\alpha) = 0 \Leftrightarrow (x - \alpha) \mid p,$$

i.e. α is a zero of p if and only if $x - \alpha$ is a factor of p.

Example 1.45 (Synthetic Division)
If the divisor is a linear factor of the form $x - x_0$, then long division can be shortened to a procedure called *synthetic division* illustrated below: Let $p(x) = a_0 + a_1x + a_2x^2 + \cdots + a_nx^n$ and $d(x) = x - x_0$,

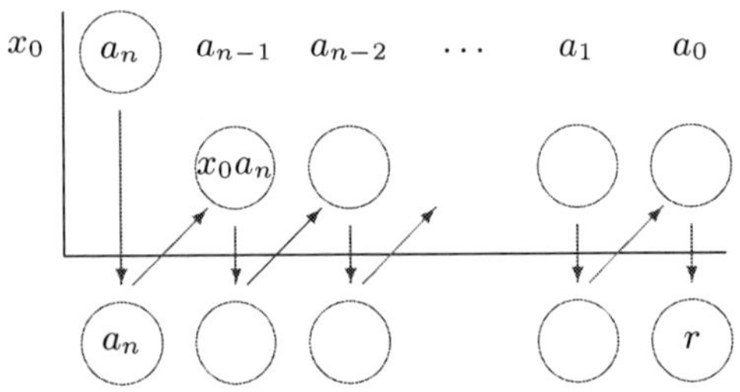

The vertical arrow represent addition of the column elements above the horizontal line, the diagonal arrows multiplication by x_0. After completion of the algorithm, the entries in the bottom row of the diagram starting with a_n are the coefficients of the quotient polynomial, the right most entry is the remainder r which coincides with the function value of p at x_0: $p(x_0) = r$.

Example 1.46 (Synthetic Division)
Divide the polynomial $p(x) = 6x^3 - 19x^2 + 16x - 7$ by the polynomial $d(x) = x - 2$ using synthetic division.

SOLUTION:

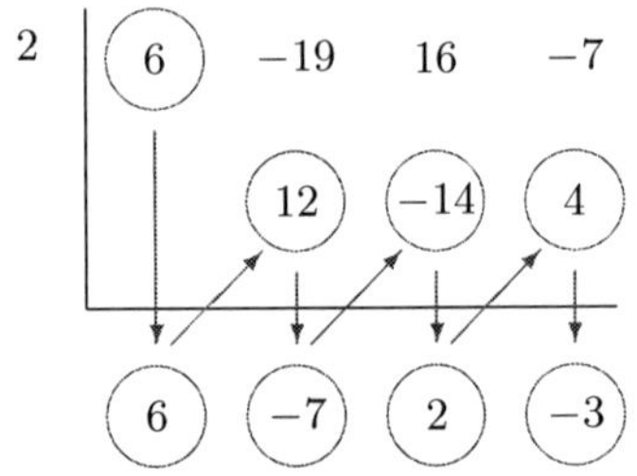

Again, we obtain

$$\overbrace{6x^3 - 19x^2 + 16x - 7}^{p(x)} = \\ = \underbrace{(x-2)}_{d(x)}\underbrace{(6x^2 - 7x + 2)}_{q(x)} + \underbrace{(-3)}_{r(x)}$$

Note that the numbers in the bottom row of the synthetic division diagram are the coefficients of the quotient polynomial in descending order, the right most coefficient is the remainder which equals the value of the poylnomial p at $x_0 = -3$. $\triangleleft$

Corollary 1.6 (Polynomials: Zeros, Factorization)
Suppose p is a polynomial over C and $\alpha \in \mathbb{C}$. Then the following three statements are equivalent:

1. $p(\alpha) = 0$, *i.e. α is a zero of p;*
2. $(x - \alpha) \mid p$, *i.e. $(x - \alpha)$ divides p;*
3. $p(x) = (x - \alpha)\, q(x)$ *where q is the quotient by division of p by $x - \alpha$.*

Example 1.47 (Factoring)
Write the polynomial

$$p(x) = x^4 - 2x^3 + 2x^2 - 2x + 1.$$

as a product of linear factors!

SOLUTION: Since $p(1) = 0$, as can be easily verified by inspection, $\alpha_1 = 1$ is a zero of p. We divide p by the linear polynomial $x - \alpha_1 = x - 1$ using synthetic division:

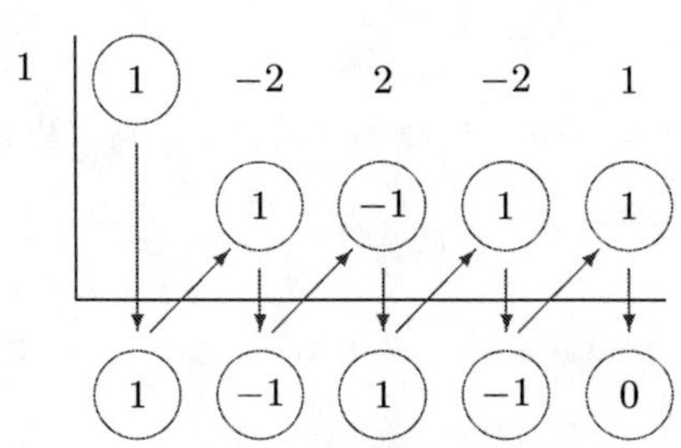

Let $q_0(x) := x^3 - x^2 + x - 1$ denote the quotient polynomial. Note that its coefficients are given by the entries of the last row in the table above. Then

$$p(x) = (x - 1)q_0(x).$$

Clearly, $q_0(1) = 0$. We devide q_0 by $(x - 1)$ using synthetic division:

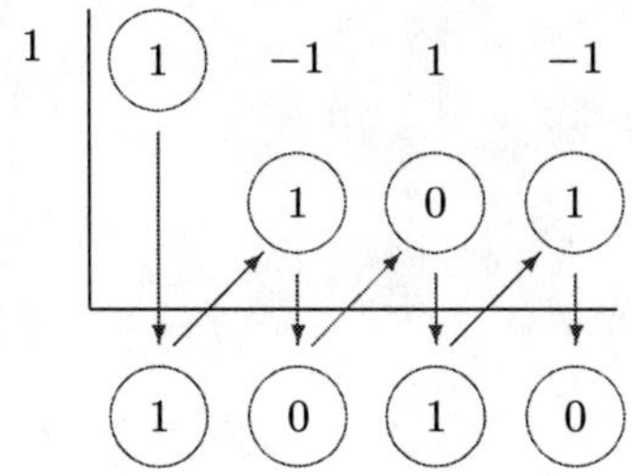

Let $q_1(x) := x^2 + 1$ denote the quotient ploynomial. Again its coefficients can be read of the last row of the synthetic division table. Hence

$$q_0(x) = (x-1)q_1(x)$$

and thus

$$\begin{aligned} p(x) &= (x-1)q_0(x) = (x-1)(x-1)q_1(x) \\ &= (x-1)^2(x^2+1) \\ &= (x-1)^2(x-i)(x+i). \end{aligned}$$

◁

Definition 1.38 (Multiplicity of a Zero/Root)
If the polynomial p is divisible by $(x-\alpha)^k$ but not by $(x-\alpha)^{k+1}$, where $k \in \mathbb{N}$, then α is called a zero or root of *multiplicity* k of p.

Example 1.48
Find the zeroes and their multiplicities of the polynomial $p(x) = x^4 - 2x^3 + 2x^2 - 2x + 1$.

SOLUTION: We have already shown in Example 1.47 that

$$p(x) = (x-1)^2(x-i)(x+i)$$

Hence, 1 is a zero of multiplicity 2, while i and $-i$ are zeros of multiplicity 1 of the polynomial p. ◁

Theorem 1.17 (The Fundamental Theorem of Algebra)
Every non-constant polynomial has at least one zero in $\mathbb{C}$.

Corollary 1.7 (Linear Factorization Theorem)
Every polynomial p *of degree* $n > 0$ *has precisely* n *zeroes. I.e. there exist numbers* $\alpha_1, \alpha_1, \ldots, \alpha_n \in \mathbb{C}$ *such that*

$$p(x) = a_n(x-\alpha_1)(x-\alpha_2)\cdots(x-\alpha_n),$$

where $a_n \in \mathbb{C}$ *is the leading coefficient of* p.

Corollary 1.8 (Factorization over $\mathbb{R}$)
Every real polynomial has a unique decomposition into a product of real linear factors and real quadratic polynomials that are "irreducible" over $\mathbb{R}$ *(i.e. that cannot be written as a product of real linear factors).*

Figure 1.22: *Carl Friedrich Gauß (1777–1855)*
The Fundamental Theorem of Algebra was first proved by Gauß.

Theorem 1.18 (Rational Zero Test)
Suppose $p(x) = a_n x^n + a_{n-1}x^{n-1} + \cdots + a_0$ is a polynomial over $\mathbb{Z}$, i.e. all coefficients $a_k \in \mathbb{Z}$ are integers, and suppose that $\alpha = r/s \in \mathbb{Q}$ is a rational zero of p. Then

$$r \mid a_0 \quad \text{and} \quad s \mid a_n,$$

i.e the numerator divides the constant term and the denominator divides the leading coefficient of p.

Example 1.49
Given the polynomial

$$p(x) = 6x^3 - 4x^2 + 3x - 2$$

Determine all rational numbers that by Theorem 1.18 could be rational zeros of p!

SOLUTION: The positive divisors of the constant term 2 are: $1, 2$; the positive divisors of the leading coefficient 6 are: $1, 2, 3, 6$. Hence the rational zeros of p are a subset of

$$\left\{\pm 1, \pm\frac{1}{2}, \pm\frac{1}{3}, \pm\frac{1}{6}, \pm\frac{2}{3}, \pm 2\right\}$$

and can be checked by inspection using synthetic division. $\triangleleft$

Definition 1.39 (Variation in Sign)
Suppose $p(x) = a_n x^n a_{n-1}x^{n-1} + \cdots + a_0$ is a polynomial over $\mathbb{R}$, i.e. $a_k \in \mathbb{R}$. We say that p has a *variation in sign* if there exists an index $k \in \{0, 1, \ldots, n-1\}$ such that $a_k \cdot a_{k+1} < 0$.

Example 1.50
Determine the variations in sign of the polynomial $p(x) = 3x^3 - 5x^2 + 6x - 4$.

SOLUTION: p has 3 variations in sign, namely:

$$3 \cdot (-5) < 0, \ (-5) \cdot 6 < 0, \ 6 \cdot (-4) < 0.$$

$\triangleleft$

Theorem 1.19 (Descartes' Rule of Signs)
Suppose $p(x) = a_n x^n + a_{n-1}x^{n-1} + \cdots + a_0$ is a real polynomial, i.e. $a_k \in \mathbb{R}$ with $a_0 \neq 0$. Then the number of positive zeroes of p equals the number of variations in sign of $p(x)$ possible reduced by an even number. Moreover, the number of negative zeroes of p coincides with the number of variations in sign of $p(-x)$ possible reduced by an even number.

Figure 1.23: *René Descartes (1596–1650).*

Example 1.51
Determine the number of positive/negative zeros of the polynomial $p(x) = 3x^3 - 5x^2 + 6x - 4$.

SOLUTION: We have seen in Example 1.50 that $p(x)$ has 3 variations in sign. Hence p has either 3 positive zeros or 1 positive zero and a pair of complex conjugate zeros. As also seen by counting the variations in sign of $p(-x) = -6x^3 - 4x^2 - 3x - 2$, p does not have any negative zeros. ◁

Theorem 1.20 (Upper and Lower Bound Rule)
Suppose $p(x) = a_n x^n + a_{n-1}x^{n-1} + \cdots + a_0$ is a real polynomial, i.e. $a_k \in \mathbb{R}$ and suppose the leading coefficient a_n is positive. Moreover, suppose that p is divided by $(x - c)$ using synthetic division. with last row $c_n = a_n, c_{n-1}, \ldots, c_0 = r$. If $c > 0$ and $c_k \geq 0$ for $k = 0, \ldots, n$, then c is an upper bound for the real zeroes of p. If $c < 0$ and the c_k have alternating signs (zero entries count as positive or negative!), then c is a lower bound for the real zeroes of p.

Example 1.52
Find the real zeroes of the polynomial

$$p(x) = 6x^3 - 4x^2 + 3x - 2.$$

SOLUTION: By the Fundamental Theorem of Algebra, p has precisely 3 zeroes. We first consider the rational zeroes of p if it has any. By the Rational Zero Test, the set Z of rational zeroes of p is contained in the set

$$\begin{aligned} Z &\subseteq \left\{ \frac{s}{t} \,\middle|\, s \mid (-2) \text{ and } t \mid 6 \right\} \\ &= \left\{ \pm 1, \pm\frac{1}{2}, \pm\frac{1}{3}, \pm\frac{1}{6}, \pm\frac{2}{3}, \pm 2 \right\}. \end{aligned}$$

Because p has three variations in sign and $p(-x) = -6x^3 - 4x^2 - 3x - 2$ has none, we can conclude with Descartes' Rule of Signs that p has no negative zeroes and either 3 positive zeroes, or 1 positiv zero (and a pair of conjugate complex zeroes). Hence

$$Z \subseteq \left\{ 1, \frac{1}{2}, \frac{1}{3}, \frac{1}{6}, \frac{2}{3}, 2 \right\}.$$

Trying $x = 1$ yields emplyoing synthetic division

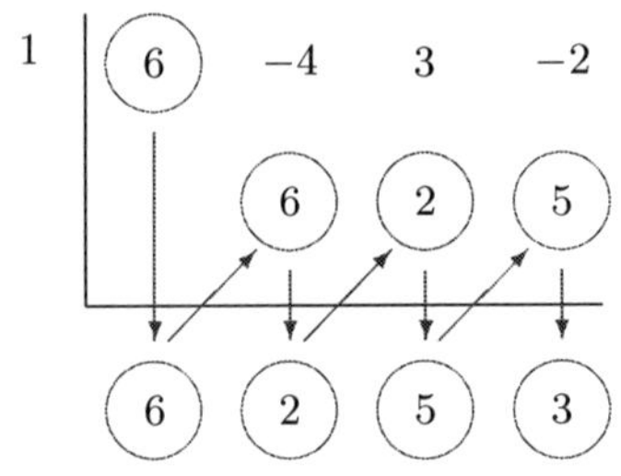

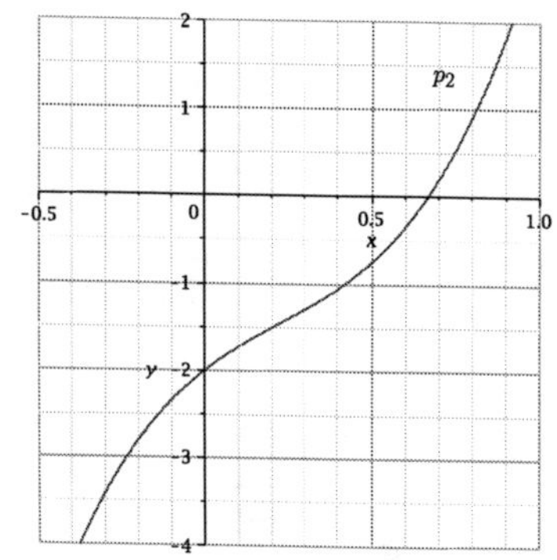

Figure 1.24:

Since the remainder $r = 3$ in the last row of the table, we see that $x = 1$ is not a zero of p. However, since the last row consists only of non-negative entries, we know from the Upper and Lower Bound Rule that 1 is an upper bound for the zeroes of p. Hence

$$Z \subseteq \left\{\frac{1}{2}, \frac{1}{3}, \frac{1}{6}, \frac{2}{3}\right\}.$$

By inspection, we find that $x = \frac{2}{3}$ is a zero of p:

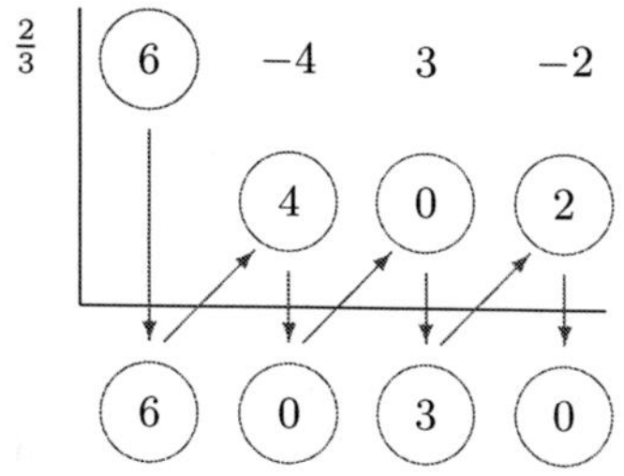

Hence $p(x) = (x - \frac{2}{3})(6x^2 + 3)$. Since $6x^2 + 3$ has no real zeroes, $x = \frac{2}{3}$ is the only real zero of the polynomial p. $\triangleleft$

Theorem 1.21 (Complex Zeros Occur in Conjugate Pairs)
Suppose p is a real polynomial (all coefficients are real numbers!), and suppose that $z = a + bi \in \mathbb{C}$ is a zero of p. Then the conjugate $\overline{z} = a - bi$ is also a zero of p.

Example 1.53
Find the zeroes of the polynomial $p(x) = x^2 + 1$.

SOLUTION: By inspection, we find that i is a zero of p. Since p is a real polynomial, also $\overline{i} = -i$ is a zero of p. Hence $p(x) = (x - i)(x + i)$. $\triangleleft$

Lemma 1.12
The Product of conjugated linear factors is a real polynomial.

PROOF: Let $a + bi \in \mathbb{C}$. Then

$$\begin{aligned} p(x) &= \Big(x - (a + bi)\Big)\Big(x - (a - bi)\Big) \\ &= \big((x - a) - bi\big)\big((x - a) + bi\big) \\ &= (x - a)^2 - (bi)^2 \\ &= x^2 - 2ax + a^2 + b^2 \end{aligned}$$

is clearly a real quadratic polynomial. $\square$

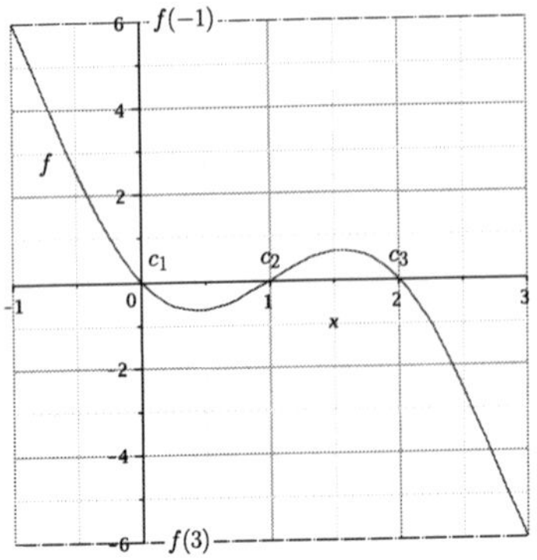

Figure 1.25:

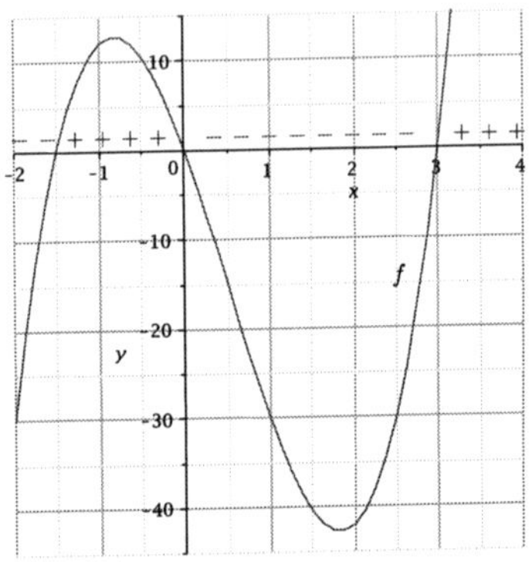

Figure 1.26:

1.1.11 Solving Inequalities

Theorem 1.22 (The Intermediate-value Theorem)
If the function f is continuous on the closed interval $[a, b]$ and d is any value between the minimum and maximum value of f then there exists at least one number $c \in [a, b]$ such that $f(c) = d$.

Corollary 1.9 (Existence Test for Zeros)
If f is continuous on the closed interval $[a, b]$ and if $f(a) \cdot f(b) < 0$ (i.e. they have different sign!) then there exists at least one number $c \in [a, b]$ such that $f(c) = 0$.

Example 1.54
Solve the inequality

$$x(6x + 9)(x - 3) > 0$$

SOLUTION: The term on the left-hand side of the inequality is a continuous function f whose (only) zeroes are $-3/2$, 0 and 3. By the corollary to the Intermediate-value Theorem, f cannot change its sign on the subintervals $(-\infty, -3/2)$, $(-3/2, 0)$, $(0, 3)$ and $(3, \infty)$, otherwise it would create further zeroes! By inspection, we find that f is positive on the intervals $(-3/2, 0)$, $(3, \infty)$ and negative on the two other intervals. Hence, the solution set of the inequality above is $S := (-3/2, 0) \cup (3, \infty)$. $\triangleleft$

1.2 Exercises

1.2.1 Fractions

Simplify the following arithmetic expressions:

1. $\left(\frac{5}{2} - \frac{1}{2}\right) \div \frac{2}{7}$ has
2. $\frac{3}{4} \cdot \left[\frac{5}{4} \div \left(\frac{3}{8} - \frac{1}{8}\right) - 3\right]$ a
3. $12 \div \frac{1}{4}$ has
4. $\left(\frac{10}{11} + \frac{6}{33} - \frac{13}{66}\right) \cdot \frac{3}{22}$ a

1.2.2 Powers, Radicals

Simplify and rationalize expressions involving exponents and radicals:

1. $\left(5^{\frac{1}{5}}\right)^{-2}$ a
2. $\left(\sqrt[4]{25}\right)^2$ a
3. $\left(25^{3/4} \cdot 4^{-3/4}\right)^2$ a
4. $\left(-\frac{125}{27}\right)^{-1/3}$ has
5. $\dfrac{\sqrt{96a^5b^{-3}}}{\sqrt{3a^{-5}b}}$ a
6. $\dfrac{x^{4/3}y^{2/3}}{(xy)^{1/3}}$ a
7. $\sqrt{\sqrt[3]{10a^7b}}$ a
8. $\sqrt[3]{16} + 3\sqrt[3]{54}$ a
9. $\dfrac{5}{\sqrt{10}}$ has
10. $\dfrac{5}{\sqrt[3]{10}}$ has
11. $\dfrac{3}{\sqrt{5} + \sqrt{6}}$ has
12. $\dfrac{3\sqrt{6} + 5\sqrt{7}}{2\sqrt{6} - 3\sqrt{7}}$ a
13. $\dfrac{3}{\sqrt[3]{5} - \sqrt[3]{6}}$ a

1.2.3 Absolute Value

Simplify the expressions involving absolute value:

1. $\left|\frac{1}{5} - \frac{1}{\sqrt{5}}\right| - |-5|$ has
2. $|-1| - |-2|$ a
3. $\dfrac{|x+2|}{x+2}$ for $x < -2$ a

1.2.4 Factorials

Simplify the expressions involving factorials:

1. $5!$ a
2. $\dfrac{8!4!}{3!5!}$ a
3. $\dfrac{4!}{3!3!}$ a

1.2.5 Domain and Range

Find the domain and range of the following functions:

1. $f(x) = |x - 5|$ a
2. $f(x) = \sqrt{x-3} - 1$ a
3. $f(x) = \sqrt{4 - x^2}$ a
4. $f(x) = \dfrac{x}{5x-3}$ a
5. $f(x) = \dfrac{\sqrt{9-x^2}}{x^2 - 3x + 2}$ has
6. $f(x) = \dfrac{1}{\sqrt{1-4x^2}}$ a

1.2.6 Composition of Functions

Find $g \circ f$, $f \circ g$ and their domains, simplify:

1. $f(x) = 2/x$, $g(x) = 2x^3 + 2x^2 + 1$ a
2. $f(x) = \sqrt[3]{x-1}$, $g(x) = x^3 + 1$ a

1.2.7 Inverse Functions

Find the inverse of the given functions if it exists. Determine domain and range:

1. Let $f(x) = \dfrac{2x+3}{3(x+1)}$ for $x \in \mathbb{R} \setminus \{-1\}$ a
2. Let $f(x) = \sqrt{4 - x^2}$ a
3. Let $f(x) = \dfrac{x+1}{x-2}$ for $x \in \mathbb{R} \setminus \{2\}$ a
4. Let $f(x) = x^2 + 9x + 5$ has
5. Let $f(x) = (4x-1)^3$ a

1.2.8 Even, odd functions

Determine whether the function is even, odd or neither:

1. $f(x) = x^6 - 2x^2 + 3$ a
2. $f(x) = x\sqrt{1-x^2}$ has
3. $f(x) = 4x^{2/3}$ ha
4. $f(x) = \tan x$ ha
5. $f(x) = \dfrac{3x^3 - 7x}{x^4 - 2x^2 - 7}$ has
6. $f(x) = x^3 - 5x$ ha
7. $f(x) = x^3 - 5$ has
8. $f(x) = 1 + x^2 \cos 2x + x$ has

1.2.9 Exponential and Logarithmic Functions

Find the domain and graph the following exponential/logarithmic functions:

1. $f(x) = \ln(x+5)$ a
2. $f(x) = 5^{x-2} + 4$ a
3. $f(x) = 1 - \ln(x-5)$ a

1.2.10 Logarithms

Manipulate logarithmic expressions:

1. Evaluate $\log_2 64$ a
2. Evaluate $\log_{32} 8$ a
3. Evaluate $\log_2 4^3$ has
4. Evaluate $\log_4 2 + \log_4 32$ a
5. Evaluate $\log_{10} 0.01$ a
6. Write as one logarithm:
   $\frac{1}{3}\left[\log_8 y + 2\log_8(y+4)\right] - \log_8(y-1)$ has

7. Simplify $4\ln 6 - 3\ln 2 + 2\ln 3 - \ln 8$ has

8. Simplify $5\ln 6 - \ln 30 + \ln \frac{1}{5}$ has

1.2.11 Complex Numbers

Transform the complex numbers into "standard form" $x + iy$:

1. $(6 - 2i)(2 - 3i)$ a

2. $(2 + 3i)^2 + (2 - 3i)^2$ a

3. $\dfrac{6 - 7i}{1 - 2i}$ has

4. $\dfrac{-3 - i}{-1 - 2i}$ has

5. $\dfrac{i}{3 - 2i} + \dfrac{2i}{3 + 8i}$ has

1.2.12 Complete the Square

Complete the square and find the vertex of the given parabola:

1. $y = x^2 - 4x + 1$ a

2. $y = 3x^2 - 6x + 12$ has

3. $y = 2x^2 + 20x + 3$ has

1.2.13 Factoring Polynomials

Solve the polynomial equations using factorization:

1. $9x^2 - 4 = 0$ has

2. $x^2 - 10x + 25 = 0$ a

3. $x^2 + 5x + 6 = 0$ a

4. $2x^2 + 5x + 3 = 0$ a

5. $25t^2 + 5t - 2 = 0$ a

6. $12x^3 - 11x^2 - 15x = 0$ ha

7. $4p^4 - 4p^3 - 8p^2 = 0$ ha

8. $4x^4 + 4x^2 + 1 = 0$ has

9. $4x^2 + 12x + 9 = 0$ a

10. $x^4 - 6x^2 + 8 = 0$ a

1.2.14 Quadratic Equations

Solve the following quadratic equations. First try factorization techniques:

1. $x^2 - x - 2 = 0$ has

2. $x^2 - x - 42 = 0$ a

3. $x^2 - 9 = 0$ a

4. $x^2 - 6x + 9 = 0$ has

5. $4m^2 - 8m + 1 = 0$ a

6. $2x^2 - 5x - 3 = 0$ has

7. $\frac{1}{3}x^2 + \frac{2}{9}x - 1 = 0$ a

8. $\dfrac{x-2}{x+2} = x - 3$ ha

9. $x^2 - 2x + 2 = 0$ has

10. $x^2 - 2x + 5 = 0$ a

11. $2x + 5\sqrt{x} = 3$ ha

12. $\left(\dfrac{1}{x+2}\right)^2 + \dfrac{4}{x+2} = 5$ ha

13. $w^4 - 5w^2 - 36 = 0$ a

1.2.15 Inequalities

Solve the following inequalities using the Intermediate-value Theorem:

1. $3x + 5 > \frac{1}{4}(x - 2)$ a
2. $2x^2 + 9x + 6 \geq x + 2$ a
3. $x(6x + 9)(x - 3) > 0$ a
4. $\dfrac{x^2 - 9}{x + 1} > 0$ a
5. $\dfrac{1}{x - 1} + \dfrac{4}{x - 6} > 0$ has
6. $\dfrac{x^2 - 4x + 3}{x^2} > 0$ has

1.2.16 Equations with Absolute Values

Solve the following equations containing absolute value terms by partitioning the domain of the equations:

1. $|y - 3| = 4$ a
2. $\left|\dfrac{3x - 1}{4}\right| = 2$ a
3. $|5y - 2| = |4y + 7|$ a

1.2.17 Inequalities with Absolute Values

Solve the following inequalities containing absolute value terms by partitioning the domain of the equation:

1. $|x - 2| < 1$ a
2. $|x - \frac{3}{2}| \geq \frac{3}{2}$ a
3. $7 < |8x - 72|$ has
4. $|x - 3| - |x + 1| \leq x$ has

1.2.18 Equations with Radicals

Solve the following equations containing radical expressions:

1. $\sqrt{17-2x}+1=x$ a

2. $\sqrt{z^2-z-7}+3=z+2$ a

3. $\sqrt{3w+4}=2+\sqrt{w}$ a

4. $\sqrt{2x+5}-\sqrt{x-1}=2$ a

5. $\sqrt[5]{x+23}=2$ a

6. $(5x-2)^{1/3}+3=0$ a

1.2.19 Exponential and Logarithmic Equations

Solve the following exponential/logarithmic equations:

1. $2^{x^2+3x}=1024$ has

2. $6(2^{3x-1})-7=9$ has

3. $e^{2x}-4e^x-5=0$ has

4. $\dfrac{400}{1+e^{-x}}=350$ a

5. $\ln\sqrt{x+2}=1$ a

6. $\ln x+\ln(x-2)=1$ a

7. $\ln\Big((2x+1)(x+2)\Big)=2\ln(x+2)$ a

8. $2\ln(x+2)-\frac{1}{2}\ln x^4=1$ has

9. $\log_4 x-\log_4(x-1)=\frac{1}{2}$ a

10. $\log_{10}4x-\log_{10}\big(12+\sqrt{x}\big)=2$ a

1.2.20 Line Equations

Find the equation of the line:

1. through $(1, 1)$, $(9, 7)$ a

2. through $(-4, 10)$, $(4, -5)$ a

3. through $(-\frac{1}{2}, 1)$, $(-\frac{5}{2}, \frac{4}{3})$ a

4. through $(1, 1)$ perpendicular to $2x + 8y = 2$ a

5. through $(2, 6)$ perpendicular to $-x + 2y = 1$ a

1.2.21 Distance and Midpoint

Find the distance and midpoint:

1. $(1, 1)$, $(9, 7)$ a

2. $(-4, 10)$, $(4, -5)$ a

3. $(-\frac{1}{2}, 1)$, $(-\frac{5}{2}, \frac{4}{3})$ a

1.2.22 Conics

Graph the following quadratic equations:

1. $x^2 - 6x + 2y + 9 = 0$ as

2. $16x^2 + 16y^2 - 16x + 24y - 3 = 0$ has

3. $x^2 + y^2 - 2x - 31 = 0$ has

4. $-9x^2 - 24x + 4y^2 - 8y + 16 = 0$ has

1.2.23 Powers and Roots, d'Moivres Theorem

1. Cube $-1 + \sqrt{3}\,i$ a

2. Cube $-1 - \sqrt{3}\,i$ has

3. Simplify i^{39} has

4. Find the square roots of $16 + 12i$ has

5. Find all 6th roots of 64. has

1.3 Hints to the Exercises

1.3.1 Fractions

Simplify the following arithmetic expressions:

1. Apply the rules for adding, subtracting, multiplying and dividing fractions!

3. Note that the integer n is identified with the fraction $\frac{n}{1}$.

1.3.2 Powers, Radicals

Simplify and rationalize expressions involving exponents and radicals:

4. Apply the laws of exponent systemically!

9. Expand the fraction by $\sqrt{10}$.

10. Expand the fraction by $10^{2/3}$.

11. Exand the fraction by $\sqrt{5} - \sqrt{6}$ and use the identity $(a + b)(a - b) = a^2 - b^2$.

1.3.3 Absolute Value

Simplify the expressions involving absolute value:

1. First determine the sign of $\left(\frac{1}{5} - \frac{1}{\sqrt{5}}\right)$.

1.3.4 Factorials

Simplify the expressions involving factorials:

1.3.5 Domain and Range

Find the domain and range of the following functions:

5. In order to determine the domain of f, eliminate those $x \in \mathbb{R}$ for which $f(x)$ is not defined (or not real). To determine the range of f, investigate the behavior of f at its poles and find its global extrema (on each subinterval of its domain) and use the continuity of f.

1.3.6 Composition of Functions

Find $g \circ f$, $f \circ g$ and their domains, simplify:

1.3.7 Inverse Functions

Find the inverse of the given functions if it exists. Determine domain and range:

4. First check whether f is injective (one-to-one). Can f be made injective by restricting it to a subset of its domain?

1.3.8 Even, odd functions

Determine whether the function is even, odd or neither:

2. Compute $f(-x)$ for an arbitrary $x \in \mathbb{R}$ and compare it with $\pm f(x)$.
3. Compute $f(-x)$ for an arbitrary $x \in \mathbb{R}$ and compare it with $\pm f(x)$.
4. Compute $f(-x)$ for an arbitrary $x \in \mathbb{R}$ and compare it with $\pm f(x)$.
5. Compute $f(-x)$ for an arbitrary $x \in \mathbb{R}$ and compare it with $\pm f(x)$.
6. Compute $f(-x)$ for an arbitrary $x \in \mathbb{R}$ and compare it with $\pm f(x)$.
7. Compute $f(-x)$ for an arbitrary $x \in \mathbb{R}$ and compare it with $\pm f(x)$.
8. Compute $f(-x)$ for an arbitrary $x \in \mathbb{R}$ and compare it with $\pm f(x)$.

1.3.9 Exponential and Logarithmic Functions

Find the domain and graph the following exponential/logarithmic functions:

1.3.10 Logarithms

Manipulate logarithmic expressions:

3. Consider the corresponding exponential equation and write 4 as power of 2.
6. Use the logarithmic rules to write the term as the logarithm of one term.
7. Use the logarithmic rules to express the logarithms in terms of $\ln 2$ and $\ln 3$.
8. Use the logarithmic rules to express the logarithms in terms of $\ln 5$ and $\ln 6$.

1.3.11 Complex Numbers

Transform the complex numbers into "standard form" $x + iy$:

3. Expand the fraction by $1 + 2i$.
4. Simplify and expand with the conjugate of the denominator.
5. Choose a common denominator and add the two fractions, then transform the sum into standard form by expanding the fraction by the conjugate of its denominator.

1.3.12 Complete the Square

Complete the square and find the vertex of the given parabola:

2. First factor out the 3
$$y = 3x^2 - 6x + 12 = 3(x^2 - 2x) + 12,$$
then use "completing the square"!

3. First factor out the 2 to obtain
$$y = 2x^2 + 20x + 3 = 2(x^2 + 10x) + 3,$$
then "complete the square"!

1.3.13 Factoring Polynomials

Solve the polynomial equations using factorization:

1. Use the formula $a^2 - b^2 = (a - b)(a + b)$ to factor the polynomial!

6. First factor out x!

7. First factor out p^2

8. Use the formulae $a^2 + 2ab + b^2 = (a + b)^2$ and $a^2 - b^2 = (a - b)(a + b)$ to factor the polynomial!

1.3.14 Quadratic Equations

Solve the following quadratic equations. First try factorization techniques:

1. Use the formula $x^2 + (a + b)x + ab = (x + a)(x + b)$ to factor the polynomial!

4. Use the Binomial formula.

6. Use the formula $\alpha_1\alpha_2 x^2 + (\alpha_1\beta_2 + \alpha_2\beta_1)x + \beta_1\beta_2 = (\alpha_1 x + \beta_1)(\alpha_2 x + \beta_2)$ to factor the polynomial!

8. First determine the domain of the equation. Then multiply by the denominator $x + 2$

9. Use the "quadratic formula" to solve the quadratic equations, since it does not have a rational zero!

11. Substitute $y = \sqrt{x}$. What is the domain of the equation?

12. First determine the domain of the equation. Then substitute $y = 1/(x + 2)$.

1.3.15 Inequalities

Solve the following inequalities using the Intermediate-value Theorem:

5. Consider the left side of the inequality as a rational function. Consider where this rational functions can change sign using the Intermediate-value Theorem.

6. Consider the left side of the inequality as a rational function and study where this function can change its sign using the Intermediate-value Theorem

1.3.16 Equations with Absolute Values

Solve the following equations containing absolute value terms by partitioning the domain of the equations:

1.3.17 Inequalities with Absolute Values

Solve the following inequalities containing absolute value terms by partitioning the domain of the equation:

3. Eliminate $|\cdot|$ by restricting the domain to the subintervals where $8x - 72 \geq 0$ and where $8x - 72 < 0$.

4. Decompose the domain $D = \mathbb{R}$ of this inequality into subdomains D_1, D_2, D_3, definded by $x - 3 \geq 3$ and $x + 1 \geq 0$, $x - 3 < 0$ and $x + 1 \geq 0$, $x - 3 < 0$ and $x + 1 < 0$ and solve the inequality first on those subdomains. Then combine the solutions.

1.3.18 Equations with Radicals

Solve the following equations containing radical expressions:

1.3.19 Exponential and Logarithmic Equations

Solve the following exponential/logarithmic equations:

1. Write 1024 as power of 2.

2. Simplify the equation and apply the respective logarithmic function to the equation.

3. First reduce the exponential equation to a quadratic equation by a suitable substitution.

8. First determine the domain of the logarithmic equation, then combine the logarithmic expressions and apply the corresponding exponential function to the compactified equation.

1.3.20 Line Equations

Find the equation of the line:

1.3.21 Distance and Midpoint

Find the distance and midpoint:

1.3.22 Conics

Graph the following quadratic equations:

2. First transform the equation into normal form by completing the square. Then interpret the resulting equation.

3. First transform the equation into normal form by completing the square. Then interpret the resulting equation.

4. First transform the equation into normal form by completing the square. Then interpret the resulting equation.

1.3.23 Powers and Roots, d'Moivres Theorem

2. First transform $-1 - \sqrt{3}\,i$ into polar form, then apply d'Moivres Theorem.

3. Reduce the power by using the fact that $i^n = i^m$ if and only if 4 divides $m - n$, i.e. $4 \mid (m - n)$.

4. Transform $16 + 12i$ into polar form and then apply d'Moivres Theorem.

5. What is the polar representation of 64?

1.4 Answers to the Exercises

1.4.1 Fractions

Simplify the following arithmetic expressions:

1. 7
2. $\frac{3}{2}$
3. 48
4. $\frac{59}{484}$

1.4.2 Powers, Radicals

Simplify and rationalize expressions involving exponents and radicals:

1. $\frac{1}{5}\left(\sqrt[5]{5}\right)^3$
2. 5
3. $\frac{125}{8}$
4. $-3/5$
5. $\dfrac{4a^5}{b^2}\sqrt{2}$
6. $x\sqrt[3]{y}$
7. $\sqrt[6]{10a^7b}$
8. $2\sqrt[3]{2}+9\sqrt[3]{2}$
9. $\frac{1}{2}\sqrt{10}$
10. $\frac{1}{2}\sqrt[3]{100}$
11. $3\left(\sqrt{6}-\sqrt{5}\right)$
12. $-\dfrac{141+19\sqrt{42}}{39}$
13. $-3\left(5^{2/3}+30^{1/3}+6^{2/3}\right)$

1.4.3 Absolute Value

Simplify the expressions involving absolute value:

1. $\frac{1}{5}\sqrt{5}-5\frac{1}{5}$
2. -1
3. -1

1.4.4 Factorials

Simplify the expressions involving factorials:

1. 120
2. 1344
3. $\frac{2}{3}$

1.4.5 Domain and Range

Find the domain and range of the following functions:

1. $D_f=\mathbb{R}$, $R_f=[0,\infty)$
2. $D_f=[3,\infty)$, $R_f=[-1,\infty)$
3. $D_f=[-2,2]$, $R_f=[0,2]$
4. $D_f=\mathbb{R}\setminus\{\frac{3}{5}\}$, $R_f=\mathbb{R}\setminus\{\frac{1}{5}\}$
5. $D_f=[-3,3]\setminus\{1,2\}$,
   $R_f=\left(-\infty,-6\sqrt{3}\right]\cup[0,\infty)$
6. $D_f=\left(-\frac{1}{2},\frac{1}{2}\right)$, $R_f=[1,\infty)$

1.4.6 Composition of Functions

Find $g \circ f$, $f \circ g$ and their domains, simplify:

1. $(g \circ f)(x) = \dfrac{x^3 + 8x + 16}{x^3}$, $(f \circ g)(x) = \dfrac{1}{x^3 + x^2 + 1/2}$
2. $(g \circ f)(x) = x$, $(f \circ g)(x) = x$

1.4.7 Inverse Functions

Find the inverse of the given functions if it exists. Determine domain and range:

1. $f^{-1}(x) = \dfrac{3x - 3}{2 - 3x}$
2. $f^{-1}(x) = \sqrt{4 - x^2}$
3. $f^{-1}(x) = \dfrac{-2x - 1}{1 - x}$
4. f is not injective (one-to-one) on $\mathbb{R}$, since, for example, $f(\frac{1}{2}) = f(-\frac{19}{2})$, as can be easily verified. Therefore, f is not invertible on $\mathbb{R}$. However, the restriction of f to $[-\frac{9}{2}, \infty)$ is a bijection from $[-\frac{9}{2}, \infty)$ onto $[-\frac{61}{4}, \infty)$ whose inverse $f^{-1} : [-\frac{61}{4}, \infty) \to [-\frac{9}{2}, \infty)$ is given by $f^{-1}(x) = -\frac{9}{2} + \sqrt{x + \frac{61}{4}}$ for all $x \in [-\frac{61}{4}, \infty)$.
5. $f^{-1}(x) = \dfrac{\sqrt[3]{x} + 1}{4}$

1.4.8 Even, odd functions

Determine whether the function is even, odd or neither:

1. even
2. odd
3. even
4. odd
5. odd
6. odd
7. neither
8. neither

1.4.9 Exponential and Logarithmic Functions

Find the domain and graph the following exponential/logarithmic functions:

1. D:$(-5, \infty)$, R:$(-\infty, \infty)$
2. D:$(-\infty, \infty)$, R:$[4, \infty)$
3. D:$(5, \infty)$, R:$(-\infty, \infty)$

1.4.10 Logarithms

Manipulate logarithmic expressions:

1. 6
2. $\frac{3}{5}$
3. 6
4. 3
5. -2
6. $\log_8 \left(\frac{y^{1/3}(y+4)^{2/3}}{y-1} \right)$
7. $-2\ln 2 + 6\ln 3$
8. $4\ln 6 - 2\ln 5$

1.4.11 Complex Numbers

Transform the complex numbers into "standard form" $x + iy$:

1. $6 - 22i$
2. -10
3. $4 + i$
4. $1 - i$
5. $\frac{62}{949} + \frac{297}{949}\, i$

1.4.12 Complete the Square

Complete the square and find the vertex of the given parabola:

1. $(x-2)^2 - 3$, vertex $V = (2, -3)$
2. $3(x-1)^2 + 9$, vertex $V = (1, 9)$
3. $2(x+5)^2 - 47$, vertex $V = (-5, -47)$

1.4.13 Factoring Polynomials

Solve the polynomial equations using factorization:

1. $x_1 = \frac{2}{3}$, $x_2 = -\frac{2}{3}$
2. $x_1 = x_2 = 5$
3. $x_1 = -2$, $x_2 = -3$
4. $x_1 = -3/2$, $x_2 = -1$
5. $t_1 = -2/5$, $t_2 = 1/5$
6. $x_1 = 0$, $x_2 = 5/3$, $x_3 = -3/4$
7. $p_1 = p_2 = 0$, $p_3 = 2$, $p_4 = -1$
8. $x_1 = x_2 = \frac{i}{2}\sqrt{2}$, $x_3 = x_4 = -\frac{i}{2}\sqrt{2}$
9. $x_1 = x_2 = -3/2$
10. $x = \pm\sqrt{2}, \pm 2$

1.4.14 Quadratic Equations

Solve the following quadratic equations. First try factorization techniques:

1. $x_1 = -1$, $x_2 = 2$
2. $x_1 = -6$, $x_2 = 7$
3. $x_1 = 3$, $x_2 = -3$
4. $x_1 = x_2 = 3$
5. $x_1 = 1 - \frac{1}{2}\sqrt{3}$, $x_2 = 1 + \frac{1}{2}\sqrt{3}$
6. $x_1 = -\frac{1}{2}$, $x_2 = 3$
7. $x_1 = -\frac{1}{3} - \frac{2}{3}\sqrt{7}$, $x_2 = \frac{1}{3} + \frac{2}{3}\sqrt{7}$
8. $x_1 = 1 - \sqrt{5}$, $x_2 = 1 + \sqrt{5}$
9. $x_1 = 1 + i$, $x_2 = 1 - i$
10. $x_1 = 1 + 2i$, $x_2 = 1 - 2i$
11. $x = 1/4$
12. $x_1 = -\frac{11}{5}$, $x_2 = -1$
13. $w_1 = -3$, $w_2 = -2i$, $w_3 = 2i$, $w_4 = 3$

1.4.15 Inequalities

Solve the following inequalities using the Intermediate-value Theorem:

1. $x > -2$
2. $-2 - \sqrt{2} \geq x \geq -2 + \sqrt{2}$
3. $-\frac{3}{2} < x < 0$, and $x > 3$
4. $-3 < x < -1$, and $x > 3$
5. $S := (1, 2) \cup (6, \infty)$
6. $S := (-\infty, 0) \cup (0, 1) \cup (3, \infty)$

1.4.16 Equations with Absolute Values

Solve the following equations containing absolute value terms by partitioning the domain of the equations:

1. $-1, 7$
2. $-7/3$, 3
3. $-5/9$, 9

1.4.17 Inequalities with Absolute Values

Solve the following inequalities containing absolute value terms by partitioning the domain of the equation:

1. $1 < x < 3$
2. $0 \geq x \geq 3$
3. $(-\infty, \frac{65}{8}) \cup (\frac{79}{8}, \infty)$
4. $[\frac{2}{3}, \infty)$

1.4.18 Equations with Radicals

Solve the following equations containing radical expressions:

1. 4
2. 8
3. $0, 4$
4. $2, 10$
5. 9
6. -5

1.4.19 Exponential and Logarithmic Equations

Solve the following exponential/logarithmic equations:

1. $x_1 = -5, x_2 = 2$
2. $x = \frac{1}{3}(4 - \log_2 3)$
3. $x_1 = \ln(5), x_2 = i\pi$
4. $x = \ln(7)$
5. $x = e^2 - 2$
6. $x = \sqrt{e+1} + 1$
7. $x = 1$
8. $x = \frac{2}{\sqrt{e}-1}$
9. $x = 2$
10. $x = \frac{25\left(5+\sqrt{73}\right)^2}{4}$

1.4.20 Line Equations

Find the equation of the line:

1. $y = 3/4x + 1/4$
2. $y = -15/8x + 5/2$
3. $y = -1/6x + 11/12$
4. $y = 4x - 3$
5. $y = -2x + 10$

1.4.21 Distance and Midpoint

Find the distance and midpoint:

1. d= 10, m= $(5, 4)$
2. d= $\sqrt{289}$, m= $(0, 5/2)$
3. d= $\sqrt{37/9}$, m= $(-3, 7/6)$

1.4.22 Conics

Graph the following quadratic equations:

1. $y = -\frac{1}{2}(x-3)^2$; parabola, opened downwardly with vertex at $(3, 0)$
2. $\left(x - \frac{1}{2}\right)^2 + \left(y + \frac{3}{4}\right)^2 = 1$; circle of radius $r = 1$ and center at $\left(\frac{1}{2}, -\frac{3}{4}\right)$.

3. $(x-1)^2+y^2=32$; circle of radius $4\sqrt{2}$ and center at $(1,0)$.

4. $\frac{(x+\frac{4}{3})^2}{\frac{28}{9}}-\frac{(y-1)^2}{7}=1$; hyperbola with asymptotes $y=1\pm\frac{3}{2}(x+\frac{4}{3})$, foci $F_1=\left(-\frac{4}{3}-\frac{1}{3}\sqrt{91},1\right)$, $F_2=\left(-\frac{4}{3}+\frac{1}{3}\sqrt{91},1\right)$, vertices $V_1=\left(-\frac{4}{3}-\frac{2}{3}\sqrt{7},1\right)$, $V_2=\left(-\frac{4}{3}+\frac{2}{3}\sqrt{7},1\right)$ and center $\left(-\frac{4}{3},1\right)$.

1.4.23 Powers and Roots, d'Moivres Theorem

1. 8

2. -8

3. -i

4. Let $\varphi=\arg(16+12i)$, then the two square roots of $16+12i$ are given by

$$z_0=2\sqrt{5}e^{\frac{\varphi}{2}i}$$
$$z_1=2\sqrt{5}e^{\frac{\varphi}{2}i+\pi i}$$

5. The six roots of 64 are

$$\begin{aligned}z_0&=2\\z_1&=1+i\sqrt{3}\\z_2&=-1+i\sqrt{3}\\z_3&=-2\\z_4&=-1-i\sqrt{3}\\z_5&=1-i\sqrt{3}.\end{aligned}$$

1.5 Solutions to Selected Exercises

1.5.1 Fractions

Simplify the following arithmetic expressions:

1. We apply the rules for adding, subtraction, multiplying and dividing fractions and obtain step by step that

$$\begin{aligned}\left(\tfrac{5}{2}-\tfrac{1}{2}\right)\div\tfrac{2}{7} &= \tfrac{5-1}{2}\div\tfrac{2}{7}=\tfrac{4}{2}\div\tfrac{2}{7}\\ &= \tfrac{2}{1}\div\tfrac{2}{7}=\tfrac{2}{1}\cdot\tfrac{7}{2}=\tfrac{2\cdot 7}{1\cdot 2}\\ &= \tfrac{14}{2}=7\end{aligned}$$

3. Here, we observe that the integers n is identified with the fraction $\frac{n}{1}$. Therefore, multiplying or dividing mixed terms can be easily reduced to working with fractions:

$$\begin{aligned}12\div\tfrac{1}{4} &= 12\cdot\tfrac{4}{1}=\tfrac{12}{1}\cdot\tfrac{4}{1}=\tfrac{12\cdot 4}{1\cdot 1}=\tfrac{48}{1}\\ &= 48\end{aligned}$$

1.5.2 Powers, Radicals

Simplify and rationalize expressions involving exponents and radicals:

4. Here, we apply the laws of exponent and obtain step by step

$$\begin{aligned}\left(-\tfrac{125}{27}\right)^{-\frac{1}{3}} &= -\left(\tfrac{125}{27}\right)^{-\frac{1}{3}}\\ &= -\left(\tfrac{27}{125}\right)^{\frac{1}{3}}=-\frac{27^{\frac{1}{3}}}{125^{\frac{1}{3}}}=-\tfrac{3}{5}\end{aligned}$$

9. We expand the fraction by $\sqrt{10}$:

$$\begin{aligned}\frac{5}{\sqrt{10}} &= \frac{5}{\sqrt{10}}\cdot 1=\frac{5}{\sqrt{10}}\cdot\frac{\sqrt{10}}{\sqrt{10}}=\frac{5\sqrt{10}}{10}\\ &= \frac{5\sqrt{10}}{10}=\tfrac{1}{2}\sqrt{10}\end{aligned}$$

10. We expand the fraction by $10^{2/3}$:

$$\begin{aligned}\frac{5}{\sqrt[3]{10}} &= \frac{5}{10^{1/3}}\cdot\frac{10^{2/3}}{10^{2/3}}=\frac{5\cdot 10^{2/3}}{10^{1/3}\cdot 10^{2/3}}\\ &= \frac{5\cdot 10^{2/3}}{10^{1/3+2/3}}=\frac{5\cdot 10^{2/3}}{10}=\tfrac{1}{2}\sqrt[3]{100}.\end{aligned}$$

11. We expand the fraction by $\sqrt{5}-\sqrt{6}$ and use the identity $(a+b)(a-b) = a^2 - b^2$:

$$\begin{aligned}\frac{3}{\sqrt{5}+\sqrt{6}} &= \frac{3\big(\sqrt{5}-\sqrt{6}\big)}{\big(\sqrt{5}+\sqrt{6}\big)\big(\sqrt{5}-\sqrt{6}\big)}\\ &= \frac{3\big(\sqrt{5}-\sqrt{6}\big)}{\big(\sqrt{5}\big)^2-\big(\sqrt{6}\big)^2}\\ &= \frac{3\big(\sqrt{5}-\sqrt{6}\big)}{5-6}\\ &= \frac{3\big(\sqrt{5}-\sqrt{6}\big)}{-1}\\ &= 3\big(\sqrt{6}-\sqrt{5}\big)\end{aligned}$$

1.5.3 Absolute Value

Simplify the expressions involving absolute value:

1. Since $-5 < 0$, we have $|-5| = -(-5) = 5$. Moreover, since $5 > \sqrt{5}$, it follows that $\frac{1}{5} < \frac{1}{\sqrt{5}}$, since the function $x \mapsto \frac{1}{x}$ is order-reversing. Hence $\frac{1}{5} - \frac{1}{\sqrt{5}} < 0$ and thus $\left|\frac{1}{5} - \frac{1}{\sqrt{5}}\right| = -\big(\frac{1}{5} - \frac{1}{\sqrt{5}}\big) = -\frac{1}{5} + \frac{1}{\sqrt{5}}$. Therefore,

$$\begin{aligned}\left|\frac{1}{5} - \frac{1}{\sqrt{5}}\right| &= -\frac{1}{5} + \frac{1}{\sqrt{5}} - 5\\ &= \frac{1}{\sqrt{5}} - 5\tfrac{1}{5} = \tfrac{1}{5}\sqrt{5} - 5\tfrac{1}{5}.\end{aligned}$$

1.5.4 Factorials

Simplify the expressions involving factorials:

1.5.5 Domain and Range

Find the domain and range of the following functions:

5. (i) Domain of f: The domain of f is the largest subset of $\mathbb{R}$ for which

$$\begin{aligned}f(x) &= \frac{\sqrt{9-x^2}}{x^2-3x+2} = \frac{\sqrt{9-x^2}}{(x-2)(x-1)}\\ &= \frac{\sqrt{9-x^2}}{\big(x-\frac{3}{2}\big)^2-\frac{1}{4}}\end{aligned}$$

is well-defined. Therefore, we have to eliminate those $x \in \mathbb{R}$ for which $0 = x^2 - 3x + 2 = (x-1)(x-2)$, i.e. $x = 1$ and $x = 2$, and confine ourselves to those $x \in \mathbb{R}$ for which $0 \leq 9 - x^2 = (3-x)(3+x)$, i.e. $x \in [-3, 3]$. Thus

$$D_f = [-3, 3] \setminus \{1, 2\}.$$

(ii) Range of f: To determine the range of f, we will first compute the range of its restrictions to each of the subdomains given by the intervals $[-3, 1)$, $(1, 2)$ and $(2, 3]$. Note that the *only* zeroes of f are the boundary points -3, and 3 and also note that f is continuous on its domain. We can therefore apply the Intermediate-value Theorem to each closed subinterval of its domain:

We begin with the subinterval $[-3, 1)$. Since $f(-3) = 0$, $f(x) \geq 0$ for all $x \in [-3, 1)$ and $\lim_{x \to 1^-} f(x) = \infty$, as can be easily verified, we conclude using the Intermediate-value Theorem that the range of the restriction of f to the interval $[-3, 1)$ is $[0, \infty)$.

We will now consider the behavior of f on the subinterval $(1, 2)$. Clearly, $\lim_{x \to 1^+} = \lim_{x \to 2^-} = -\infty$ and $f(x) < 0$ for all $x \in (1, 2)$. It can be easily verified that f attains its global maximum on the subinterval $(1, 2)$ where the denominator of f which is given by the parabola $\left(x - \frac{3}{2}\right)^2 - \frac{1}{4}$ has its smallest value which is precisely at $\frac{3}{2}$ where the parabola has its vertex. The corresponding function value of f is

$$f\left(\tfrac{3}{2}\right) = \frac{\sqrt{9 - \left(\frac{3}{2}\right)^2}}{-\frac{1}{4}} = -6\sqrt{3}.$$

Hence, again by the Intermediate-value Theorem, the range of the restriction of f to the interval $(1, 2)$ is given by the interval $[-6\sqrt{3}, -\infty)$.

Finally, consider the interval $(2, 3]$. Clearly, $f(x) \geq 0$ for all $x \in (2, 3]$, $f(3) = 0$ and $\lim_{x \to 2^+} = \infty$, as can be easily verified. Hence, again by the Intermediate-value Theorem, the range of the restriction of f to the interval $(2, 3]$ is $[0, \infty)$.

The range of f on its total domain $D_f = [-3, 1) \cup (1, 2) \cup (2, 3]$ is therefore the union of the ranges of the restrictions of f. Hence

$$\begin{aligned}\text{range}(f) &= [0, \infty) \cup [-6\sqrt{3}, -\infty) \cup [0, \infty) \\ &= [-6\sqrt{3}, -\infty) \cup [0, \infty).\end{aligned}$$

1.5.6 Composition of Functions

Find $g \circ f$, $f \circ g$ and their domains, simplify:

1.5.7 Inverse Functions

Find the inverse of the given functions if it exists. Determine domain and range:

4. Clearly, the function $f : \mathbb{R} \to \mathbb{R}$ defined by

$$f(x) = x^2 + 9x + 5 = \left(x + \tfrac{9}{2}\right)^2 - \tfrac{61}{4}$$

(note that its graph is a parabola!) is not injective (one-to-one), since, for example,

$$\begin{aligned} f\left(\tfrac{1}{2}\right) &= \left(\tfrac{1}{2} + \tfrac{9}{2}\right)^2 - \tfrac{61}{4} \\ &= 5^2 - \tfrac{61}{4} = (-5)^2 - \tfrac{61}{4} \\ &= \left(-\tfrac{19}{2} + \tfrac{9}{2}\right)^2 - \tfrac{61}{4} = f\left(-\tfrac{19}{2}\right). \end{aligned}$$

Therefore, f is not invertible on $\mathbb{R}$. However, we can, for example, restrict it to the subinterval $[-\tfrac{9}{2}, \infty)$. This restriction, which we will also denote by f, constitutes a bijection $f : [-\tfrac{9}{2}, \infty) \to [-\tfrac{61}{4}, \infty)$ as can be easily verified. We will determine its inverse: To this end, first set $y = \left(x + \tfrac{9}{2}\right)^2 - \tfrac{61}{4}$. Next interchange x and y, which leads to the equation $x = \left(y + \tfrac{9}{2}\right)^2 - \tfrac{61}{4}$. Then solve for y: Taking the square root of

$$\left(y + \tfrac{9}{2}\right)^2 = x + \tfrac{61}{4}$$

yields

$$\left|y + \tfrac{9}{2}\right| = \sqrt{x + \tfrac{61}{4}}$$

and

$$y + \tfrac{9}{2} = \sqrt{x + \tfrac{61}{4}}$$

since $y \in [-\tfrac{9}{2}, \infty)$, implies $\left|y + \tfrac{9}{2}\right| = y + \tfrac{9}{2}$. Thus

$$y = -\tfrac{9}{2} + \sqrt{x + \tfrac{61}{4}}.$$

Hence, the inverse function

$$f^{-1} : [-\tfrac{61}{4}, \infty) \to [-\tfrac{9}{2}, \infty)$$

is given by

$$f^{-1}(x) = -\tfrac{9}{2} + \sqrt{x + \tfrac{61}{4}}$$

for $x \in [-\tfrac{61}{4}, \infty)$.

We conclude by showing that f^{-1} is indeed inverse to f on $[-\frac{9}{2},\infty)$: Let $x \in [-\frac{9}{2},\infty)$ then

$$\begin{aligned}
f^{-1}(f(x)) &= f^{-1}\Big(\big(x+\tfrac{9}{2}\big)^2 - \tfrac{61}{4}\Big)\\
&= -\tfrac{9}{2} + \sqrt{\big(x+\tfrac{9}{2}\big)^2 - \tfrac{61}{4} + \tfrac{61}{4}}\\
&= -\tfrac{9}{2} + \big|x+\tfrac{9}{2}\big|\\
&= -\tfrac{9}{2} + x + \tfrac{9}{2}\\
&= x.
\end{aligned}$$

Conversely, let $x \in [-\frac{61}{4},\infty)$ then

$$\begin{aligned}
f(f^{-1}(x)) &= f\Big(-\tfrac{9}{2} + \sqrt{x+\tfrac{61}{4}}\Big)\\
&= \Big(-\tfrac{9}{2} + \sqrt{x+\tfrac{61}{4}} + \tfrac{9}{2}\Big)^2 - \tfrac{61}{4}\\
&= x + \tfrac{61}{4} - \tfrac{61}{4}\\
&= x.
\end{aligned}$$

1.5.8 Even, odd functions

Determine whether the function is even, odd or neither:

2. We compute $f(-x)$ and check whether it equals $f(x)$ or $-f(x)$ or neither:

$$\begin{aligned}
f(-x) &= (-x)\sqrt{1-(-x)^2}\\
&= \Big(-x\sqrt{1-x^2}\Big)\\
&= -f(x)
\end{aligned}$$

Since, $f(-x) = -f(x)$ for all $x \in \mathbb{R}$, f is odd on $\mathbb{R}$.

5. We compute $f(-x)$ and check whether it equals $f(x)$ or $-f(x)$ or neither:

$$\begin{aligned}
f(-x) &= \frac{3(-x)^3 - 7(-x)}{(-x)^4 - 2(-x)^2 - 7}\\
&= \frac{-3x^3 + 7}{x^4 - 2x^2 - 7}\\
&= -\frac{3x^3 - 7}{x^4 - 2x^2 - 7}\\
&= -f(x).
\end{aligned}$$

Since $f(-x) = -f(x)$ for all $x \in \mathbb{R}$, f is odd on $\mathbb{R}$.

7. We compute $f(-x)$ and check whether it equals $f(x)$ or $-f(x)$ or neither:

$$f(-x) = (-x)^3 - 5 = -x^3 - 5$$

which is clearly not equal to either $f(x)$ or $-f(x)$ for all $x \in \mathbb{R}$: Choose for example $x = 1$. Then, $f(-1) = -6$ while $f(1) = -4$. Hence, f is neither odd nor even on $\mathbb{R}$.

8. We compute $f(-x)$ and check whether it equals $f(x)$ or $-f(x)$ or neither:

$$\begin{aligned} f(-x) &= 1 + (-x)^2 \cos\big(2(-x)\big) + (-x) \\ &= 1 + x^2 \cos(2x) - x \end{aligned}$$

Again, we see that $f(-x)$ is not equal to either $f(x)$ or $f(-x)$ for all $x \in \mathbb{R}$: Choose, for instance, $x = \frac{\pi}{4}$, then $f\left(-\frac{\pi}{4}\right) = 1 - \frac{\pi}{4}$, while $f\left(\frac{\pi}{4}\right) = 1 + \frac{\pi}{4}$. Hence f is neither odd nor even on $\mathbb{R}$.

1.5.9 Exponential and Logarithmic Functions

Find the domain and graph the following exponential/logarithmic functions:

1.5.10 Logarithms

Manipulate logarithmic expressions:

3. Recall that the base-2 exponential function $\exp_2$ is the inverse function of $\log_2$. Applying $\exp_2$ on both sides of the equation, we see that $\log_2 4^3 = y \Leftrightarrow 2^y = 4^3 = \left(2^2\right)^3 = 2^6$. Since $\exp_2$ is one-to-one, it follows that $y = 6$.

6. Using the logarithmic rules, we obtain

$$\begin{aligned} &\tfrac{1}{3}\left[\log_8 y + 2\log_8(y+4)\right] - \log_8(y-1) \\ &= \tfrac{1}{3}\left[\log_8 y + \log_8(y+4)^2\right] - \log_8(y-1) \\ &= \tfrac{1}{3}\log_8\big(y(y+4)^2\big) - \log_8(y-1) \\ &= \log_8\left(y(y+4)^2\right)^{1/3} - \log_8(y-1) \\ &= \log_8\left(\frac{y^{1/3}(y+4)^{2/3}}{y-1}\right). \end{aligned}$$

7. We apply the well-known logarithmic rules step by step and obtain

$$\begin{aligned}
&4\ln 6 - 3\ln 2 + 2\ln 3 - \ln 8 = \\
&\quad = 4\ln(2\cdot 3) - 3\ln 2 + 2\ln 3 - \ln 2^3 \\
&\quad = 4\big(\ln 2 + \ln 3\big) - 3\ln 2 + 2\ln 3 - 3\ln 2 \\
&\quad = 4\ln 2 + 4\ln 3 - 3\ln 2 + 2\ln 3 - 3\ln 2 \\
&\quad = (4 - 3 - 3)\ln 2 + (4 + 2)\ln 3 \\
&\quad = -2\ln 2 + 6\ln 3.
\end{aligned}$$

8. We apply the logarithmic rules and obtain

$$\begin{aligned}
&5\ln 6 - \ln 30 + \ln \tfrac{1}{5} = \\
&\quad = 5\ln 6 - \ln(5\cdot 6) - \ln 5 \\
&\quad = 5\ln 6 - \big(\ln 5 + \ln 6\big) - \ln 5 \\
&\quad = 5\ln 6 - \ln 5 - \ln 6 - \ln 5 \\
&\quad = 4\ln 6 - 2\ln 5.
\end{aligned}$$

1.5.11 Complex Numbers

Transform the complex numbers into "standard form" $x + iy$:

3. We expand the fraction by the conjugate of the doniminator $1 - 2i$:

$$\begin{aligned}
\frac{6-7i}{1-2i} &= \frac{6-7i}{1-2i}\cdot\frac{1+2i}{1+2i} \\
&= \frac{6\cdot 1 + (-7i)(2i) + 6\cdot 2i + (-7i)1}{1-(2i)^2} \\
&= \frac{6 + 14 + i(12-7)}{1+4} \\
&= \frac{20+5i}{5} = 4 + i.
\end{aligned}$$

4. We first simplify the fraction and then expand by the conjugate of the denominator:

$$\begin{aligned}
\frac{-3-i}{-1-2i} = \frac{-(3+i)}{-(1+2i)} &= \frac{3+i}{1+2i} \\
&= \frac{3+i}{1+2i}\cdot\frac{1-2i}{1-2i} \\
&= \frac{3+2+i(-6+1)}{1^2-(2i)^2} \\
&= \frac{5-5i}{5} = 1 - i.
\end{aligned}$$

5. We first add the two fractions and then transform the sum into standard form by expanding by the conjugate of its denominator:

$$\begin{aligned}\frac{i}{3-2i}+\frac{2i}{3+8i} &= \frac{i(3+8i)+2i(3-2i)}{(3-2i)(3+8i)}\\ &= \frac{3i-8+6i+4}{9+16+i(24-6)}\\ &= \frac{-4+9i}{25+18i}\\ &= \frac{-4+9i}{25+18i}\cdot\frac{25-18i}{25-18i}\\ &= \frac{-100+162+i(72+225)}{625+324}\\ &= \frac{62+297i}{949}\\ &= \frac{62}{949}+\frac{297}{949}i\end{aligned}$$

1.5.12 Complete the Square

Complete the square and find the vertex of the given parabola:

2.

$$\begin{aligned}y = 3x^2-6x+12 &= 3(x^2-2x)+12\\ &= 3\big((x-1)^2-(-1)^2\big)+12\\ &= 3(x-1)^2-3+12\\ &= 3(x-1)^2+9.\end{aligned}$$

The vertex of a parabola is its local extremum, which is clearly attained when the term $(x-1)^2$ has its minimum, hence $V=(1,9)$.

3.

$$\begin{aligned}y = 2x^2+20x+3 &= 2(x^2+10x)+3\\ &= 2\big((x+5)^2-25\big)+3\\ &= 2(x+5)^2-47,\end{aligned}$$

hence the vertex is $V=(-5,-47)$.

1.5.13 Factoring Polynomials

Solve the polynomial equations using factorization:

1. We us the formula $a^2-b^2=(a-b)(a+b)$ to factor the polynomial:

$$\begin{aligned}0 = 9x^2-4 &= (3x)^2-2^2\\ &= (3x-2)(3x+2).\end{aligned}$$

Hence, $3x - 2 = 0$ or $3x + 2 = 0$ and thus $x = \frac{2}{3}$ or $x = -\frac{2}{3}$.

8. We use the formulae $a^2+2ab+b^2 = (a+b)^2$ and $a^2-b^2 = (a-b)(a+b)$ to factor the polynomial:

$$\begin{aligned} 0 &= 4x^4 + 4x^2 + 1 \\ &= \left(2x^2\right)^2 + 2(2x^2)\cdot 1 + 1^2 \\ &= \left(2x^2+1\right)^2 = \left(2(x^2+\tfrac{1}{2})\right)^2 \\ &= 4(x^2+\tfrac{1}{2})^2 \\ &= 4\left(x+\tfrac{i}{2}\sqrt{2}\right)^2\left(x-\tfrac{i}{2}\sqrt{2}\right)^2. \end{aligned}$$

Hence, $x = -\frac{i}{2}\sqrt{2}$ or $x = \frac{i}{2}\sqrt{2}$ are double roots of the given polynomial.

1.5.14 Quadratic Equations

Solve the following quadratic equations. First try factorization techniques:

1. We use the formula $x^2 + (a+b)x + ab = (x+a)(x+b)$ to factor the polynomial:

$$0 = x^2 - x - 2 = (x-2)(x+1),$$

hence, $x = 2$ or $x = -1$.

4. Using the Binomial formula, we can easily factor the given polynomial

$$0 = x^2 - 6x + 9 = (x-3)^2,$$

which implies that $x_1 = x_2 = 3$.

6. We use the formula $\alpha_1\alpha_2 x^2 + (\alpha_1\beta_2 + \alpha_2\beta_1)x + \beta_1\beta_2 = (\alpha_1 x + \beta_1)(\alpha_2 x + \beta_2)$ to factor the polynomial:

$$0 = 2x^2 - 5x - 3 = (2x+1)(x-3),$$

hence, $2x+1 = 0$ or $x - 3 = 0$ and thus $x = -\frac{1}{2}$ or $x = 3$.

9. We use the "quadratic formula" to solve the quadratic equations, since it does not have a rational zero:

$$\begin{aligned} x &= \frac{-(-2) \pm \sqrt{(-2)^2 - 4\cdot 1\cdot 2}}{2\cdot 1} \\ &= \frac{2 \pm \sqrt{-4}}{2} = \frac{2 \pm 2i}{2} = 1 \pm i. \end{aligned}$$

1.5.15 Inequalities

Solve the following inequalities using the Intermediate-value Theorem:

5. We consider the left side of the inequality as a rational function and set

$$\begin{aligned} r(x) &= \frac{1}{x-1} + \frac{4}{x-6} = \frac{5x-10}{(x-1)(x-6)} \\ &= \frac{5(x-2)}{(x-1)(x-6)} \end{aligned}$$

Clearly, the domain of f is $D_r := \mathbb{R} \setminus \{1, 6\}$. As a rational function r is continuous on its domain D_r and by the Intermediate-value Theorem of Calculus can only change its sign (change from positive values to negative or vice versa) at its only zero $x_0 = 2$ or at its singularities $x_1 = 1$ and $x_2 = 6$.

In order to determine the set S of all such points $x \in D_r$ for which $r(x) > 0$, it suffices to determine the sign of r at one "test point" of each of the subintervals $(-\infty, 1)$, $(1, 2)$, $(2, 6)$ and $(6, \infty)$ of D_r obtained by removing the zeroes of the numberator and denominator of r from $\mathbb{R}$. We choose the points $0, \frac{3}{2}, 3, 7$ as test points on those intervals and find (please verify!) that $r(0) < 0$, $r(\frac{3}{2}) > 0$, $r(3) < 0$ and $r(7) > 0$. Hence, we conclude that

$$r(x) \begin{cases} < 0, & x \in (-\infty, 1) \\ > 0, & x \in (1, 2) \\ < 0, & x \in (2, 6) \\ > 0, & x \in (6, \infty). \end{cases}$$

Thus, the solution set of the given inequality is $S := (1, 2) \cup (6, \infty)$.

6. The rational function

$$r(x) := \frac{x^2 - 4x + 3}{x^2} = \frac{(x-1)(x-3)}{x^2}$$

is continuous on $\mathbb{R} \setminus \{0\}$. Hence, considering that $r(x) = 0 \Leftrightarrow x = 1$ or $x = 3$ and applying the Intermediate-value Theorem, we find that r can only "change sign" at $0, 1, 3$. In order to determine the sign of r on the intervals $(-\infty, 0)$, $(0, 1)$, $(1, 3)$ and $(3, \infty)$, we pick test points in each of these intervals, say, -1, $\frac{1}{2}$, 2 and 4 and check the sign of r at those points. We find that $r(-1) > 0$, r($\frac{1}{2}$) > 0, $r(2) < 0$ and $r(4) > 0$. Hence, we conclude by the Intermediate-value Theorem that

$$r(x) \begin{cases} > 0 & x \in (-\infty, 0) \\ > 0 & x \in (0, 1) \\ < 0 & x \in (1, 3) \\ > 0 & x \in (3, \infty) \end{cases}$$

Hence, the solution set of the given inequality is
$S := (-\infty, 0) \cup (0, 1) \cup (3, \infty)$.

1.5.16 Equations with Absolute Values

Solve the following equations containing absolute value terms by partitioning the domain of the equations:

1.5.17 Inequalities with Absolute Values

Solve the following inequalities containing absolute value terms by partitioning the domain of the equation:

3. We are asked to find all $x \in \mathbb{R}$ that satisfy the inequality $7 < |8x - 72|$. First note that

$$|8x - 72| = \begin{cases} 8x - 72, & 8x - 72 \geq 0 \\ -(8x - 72), & 8x - 72 < 0 \end{cases}$$
$$= \begin{cases} 8x - 72, & x \geq 9 \\ -8x + 72, & x < 9. \end{cases}$$

We will therefore solve the inequality on the subdomains $D_1 := [9, \infty)$ and $D_2 := (-\infty, 9)$, where $|8x - 72|$ can be expressed without $|\cdot|$, separately:

(i) First, let $x \in D_1 = [9, \infty)$, then $|8x - 72| = 8x - 72$ and

$$7 < |8x - 72|$$

is equivalent to the following inequalities

$$\begin{aligned} 7 &< 8x - 72 \\ 79 &< 8x \\ \tfrac{79}{8} &< x, \end{aligned}$$

which implies that the solution set of the given inequality on D_1 is $S_1 := (\frac{79}{8}, \infty)$.

(ii) Next, let $x \in D_2 = (-\infty, 9)$, then $|8x - 72| = -8x + 72$ and similar as in the preceeding case

$$7 < |8x - 72|$$

is equivalent to the inequalities

$$\begin{aligned} 7 &< -8x + 72 \\ 8x &< 65 \\ x &< \tfrac{65}{8}, \end{aligned}$$

from which follows that the solution set of the given inequality on D_2 is $S_2 := (-\infty, \frac{65}{8})$.

The solution set of the inequality $7 < |8x - 72|$ on $\mathbb{R}$ is therefore the union of the solution sets on the subdomains, i.e.

$$S := S_1 \cup S_2 = (\tfrac{79}{8}, \infty) \cup (-\infty, \tfrac{65}{8}).$$

4. We are to determine all $x \in \mathbb{R}$ satisfying the inequality $|x-3|-|x+1| \leq x$. To this end, we first observe that

$$\begin{aligned} |x-3| &= \begin{cases} x-3, & x-3 \geq 0 \\ -(x-3), & x-3 < 0 \end{cases} \\ &= \begin{cases} x-3, & x \geq 3 \\ -x+3, & x < 3 \end{cases} \end{aligned}$$

and

$$\begin{aligned} |x+1| &= \begin{cases} x+1, & x+1 \geq 0 \\ -(x+1), & x+1 < 0 \end{cases} \\ &= \begin{cases} x+1, & x \geq -1 \\ -x-1, & x < -1 \end{cases} \end{aligned}$$

Therefore, we will subdivide the domain $\mathbb{R}$ into the subregions $D_1; = [3, \infty)$, $D_2 := [-1, 3)$ and $D_3 := (-\infty, -1)$ on which the terms $|x - 3|$ and $|x + 1|$ can be simultaneously replaced by absolute value free expressions. On these subdomains, the given inequality can be rewritten without absolute value expressions:

(i) Let $x \in D_1 = [3, \infty)$, then $|x - 3| = x - 3$ and $|x + 1| = x + 1$ and

$$|x-3| - |x+1| \leq x$$

in turn is equivalent to

$$\begin{aligned} x - 3 - (x+1) &\leq x \\ x - 3 - x - 1 &\leq x \end{aligned}$$

and thus to

$$-4 \leq x,$$

which implies that the solution set of the given inequality on D_1 is $S_1 := D_1 = [3, \infty)$.

(ii) Next, let $x \in D_2 = [-1, 3)$, then $|x-3| = -x+3$ and $|x+1| = x+1$ and

$$|x-3| - |x+1| \le x$$

is equivalent to the following inequalities

$$\begin{aligned} -x+3-(x+1) &\le x \\ -2x+2 &\le x \\ x &\le 3x - \tfrac{2}{3} \qquad\qquad \le x, \end{aligned}$$

from which follows that the solution set of the given inequality on D_2 is $S_2 := [\frac{2}{3}, 3)$.

(iii) Finally, let $x \in D_3 = (-\infty, -1)$, then $|x-3| = -x+3$ and $|x+1| = -x-1$. Then

$$|x-3| - |x+1| \le x$$

is again equivalent to the following inequalities

$$\begin{aligned} -x+3-(-x-1) &\le x \\ -x+3+x+1 &\le x \\ 4 &\le x, \end{aligned}$$

from which follows that the solution set of the given inequality on D_3 is empty, i.e. $S_3 := \emptyset$.

The solution set of $|x-3| - |x+1| \le x$ on $\mathbb{R}$ is therefore, the union

$$\begin{aligned} S &= S_1 \cup S_2 \cup S_3 \\ &= [3, \infty) \cup [\tfrac{2}{3}, 3) \cup \emptyset \\ &= [\tfrac{2}{3}, \infty) \end{aligned}$$

1.5.18 Equations with Radicals

Solve the following equations containing radical expressions:

1.5.19 Exponential and Logarithmic Equations

Solve the following exponential/logarithmic equations:

1. Since the (base-2) exponential function is one-to-one, the identity $2^{x^2+3x} = 1024 = 2^{10}$ implies

$$10 = x^2 + 3x$$

and therefore

$$0 = x^2 + 3x - 10 = (x + 5)(x - 2),$$

hence $x = -5$ or $x = 2$.

2. Clearly, the exponential equation

$$6\big(2^{3x-1}\big) - 7 = 9$$

is equivalent to

$$2^{3x-1} = \frac{16}{2 \cdot 3} = \frac{2^3}{3}$$

and thus to

$$2^{3x-4} = \frac{1}{3}.$$

Applying the logarithmic function $\log_2$ to this equations first yields

$$3x - 4 = -\log_2 3$$

and then

$$x = \tfrac{1}{3}\big(4 - \log_2 3\big).$$

3. The exponential equation

$$0 = e^{2x} - 4e^x - 5$$

can be reduced to a quadratic equation by the substitution $y := e^x$:

$$0 = y^2 - 4y - 5 = (y - 5)(y + 1).$$

The solutions of the quadratic equation are $y = 5$ and $y = -1$. We have to resubstitute the permissible values of y back into the equation $y = e^x$. The first solution $y = 5$, produces the equation $5 = e^x$ which implies that $x = \ln 5$, while the second solution $y = -1$ yields the equation $-1 = e^x$ which has the complex solution $x = i\pi$.

8. First, we observe that the given equation has domain $(0, \infty)$. Thus, if $x \in (0, \infty)$, we can conclude using the logarithmic rules that

$$\begin{aligned}
1 &= 2\ln(x+2) - \frac{1}{2}\ln x^4 \\
&= 2\ln(x+2) - 2\ln x \\
&= 2\Big(\ln(x+2) - \ln x\Big) \\
&= 2\ln\left(\frac{x+2}{x}\right).
\end{aligned}$$

Dividing the equation by 2, yields

$$\frac{1}{2} = \ln\left(\frac{x+2}{x}\right)$$

Applying the natural exponential function exp (which is the inverse function of ln) on both sides of the equation yields

$$e^{\frac{1}{2}} = \exp\left(\ln\left(\frac{x+2}{x}\right)\right) = \frac{x+2}{x}$$

from which by multiplying the equation by x first follows that

$$x\sqrt{e} = x + 2$$

and then

$$x(\sqrt{e} - 1) = 2,$$

hence,

$$x = \frac{2}{\sqrt{e} - 1}.$$

1.5.20 Line Equations

Find the equation of the line:

1.5.21 Distance and Midpoint

Find the distance and midpoint:

1.5.22 Conics

Graph the following quadratic equations:

1. We transform the equation

$$x^2 - 6x + 2y + 9 = 0$$

into normal form by "completing the square"

$$\begin{aligned}(x-3)^2 - 9 + 2y + 9 &= 0\\ (x-3)^2 + 2y &= 0,\end{aligned}$$

hence, $y = -\frac{1}{2}(x-3)^2$, which is the equation of a parabola opened downwardly with vertex at $(3, 0)$.

2. We transform the equation

$$16x^2 + 16y^2 - 16x + 24y - 3 = 0$$

into normal form by "completing the square"

$$\begin{aligned} 16(x^2 - x) + 16(y^2 + \tfrac{3}{2}y) &= 3 \\ 16\big((x - \tfrac{1}{2})^2 - \tfrac{1}{4}\big) + 16\big((y + \tfrac{3}{4})^2 - \tfrac{9}{16}\big) &= 3 \\ 16\big(x - \tfrac{1}{2}\big)^2 + 16\big(y + \tfrac{3}{4}\big)^2 &= 16. \end{aligned}$$

Division by 16 yields

$$\big(x - \tfrac{1}{2}\big)^2 + \big(y + \tfrac{3}{4}\big)^2 = 1,$$

which is the equation of a circle of radius 1 and center $\big(\tfrac{1}{2}, -\tfrac{3}{4}\big)$.

3. We transform the equation

$$x^2 + y^2 - 2x - 31 = 0$$

into normal form by "completing the square"

$$\begin{aligned} (x^2 - 2x) + y^2 - 31 &= 0 \\ (x - 1)^2 - 1 + y^2 - 31 &= 0 \\ (x - 1)^2 + y^2 &= 32, \end{aligned}$$

which is the equation of a circle of radius $4\sqrt{2}$ and center $(1, 0)$.

4. We first transform the equation

$$-9x^2 - 24x + 4y^2 - 8y + 16 = 0$$

into normal form by "completing the square"

$$\begin{aligned} -9(x^2 + \tfrac{8}{3}x) + 4(y^2 - 2y) &= -16 \\ -9\big((x + \tfrac{4}{3})^2 - \tfrac{16}{9}\big) + 4\big((y - 1)^2 - 1\big)\big) &= -16 \\ -9\big(x + \tfrac{4}{3}\big)^2 + 16 + 4(y - 1)^2 - 4 &= -16 \\ -9\big(x + \tfrac{4}{3}\big)^2 + 4(y - 1)^2 &= -28 \\ \frac{\big(x + \frac{4}{3}\big)^2}{\frac{28}{9}} - \frac{(y - 1)^2}{7} &= 1. \end{aligned}$$

Clearly,

$$\frac{\big(x + \frac{4}{3}\big)^2}{\frac{28}{9}} - \frac{(y - 1)^2}{7} = 1. \tag{1.14}$$

is the equation of a hyperbola obtained from the hyperbola

$$\frac{x^2}{\frac{28}{9}} - \frac{y^2}{7} = 1, \tag{1.15}$$

which is in standard position, by a shift by the vector $\left(-\frac{4}{3}, 1\right)$.

We will first determine the asymptotes, vertices, center and foci of the hyperbola (1.15). Recall that the asymptotes of (1.15) can be easily determined from the modified equation

$$\frac{x^2}{\frac{28}{9}} - \frac{y^2}{7} = 0$$

by solving for y. Clearly, $y^2 = \frac{7 \cdot 9}{28} x^2 = \frac{9}{4} x^2$. Therefore, the asymptotes of the hyperbola (1.15) are given by the equations

$$y = \pm \tfrac{3}{2} x.$$

Moreover, the foci $(\pm c, 0)$ can be determined from the equation $c^2 = a^2 + b^2$, where $a^2 = \frac{28}{9}$ and $b^2 = 7$. Since $c = \sqrt{\frac{28}{9} + 7} = \frac{1}{3}\sqrt{28 + 63} = \frac{1}{3}\sqrt{91}$, the foci of (1.15) are given by

$$F_1 = \left(-\tfrac{1}{3}\sqrt{91}, 0\right), \quad F_2 = \left(\tfrac{1}{3}\sqrt{91}, 0\right).$$

The vertices $(\pm a, 0)$ are obtained from $a = \sqrt{\frac{28}{9}} = \frac{2}{3}\sqrt{7}$ and are

$$V_1 = \left(-\tfrac{2}{3}\sqrt{7}, 0\right), \quad V_2 = \left(\tfrac{2}{3}\sqrt{7}, 0\right).$$

Clearly the center C of the hyperbola (1.15) is at $(0, 0)$ and constitutes the midpoint of the transverse axis determined by the foci F_1 and F_2.

In order to find the asymptotes, foci, vertices and center of the hyperbola (1.14), we only have to shift the corresponding equations or points, respectively.

Thus the asymptotes of (1.14) are

$$y - 1 = \pm \frac{3}{2}\left(x + \frac{4}{3}\right),$$

the foci

$$\begin{aligned} \tilde{F}_1 &= \left(-\tfrac{1}{3}\sqrt{91} - \tfrac{4}{3}, 0 + 1\right) \\ &= \left(-\tfrac{4}{3} - \tfrac{1}{3}\sqrt{91}, 1\right) \\ \tilde{F}_2 &= \left(\tfrac{1}{3}\sqrt{91} - \tfrac{4}{3}, 0 + 1\right) \\ &= \left(-\tfrac{4}{3} + \tfrac{1}{3}\sqrt{91}, 1\right) \end{aligned}$$

the vertices

$$\begin{aligned}\tilde{V}_1 &= \left(-\tfrac{2}{3}\sqrt{7} - \tfrac{4}{3}, 0 + 1\right)\\ &= \left(-\tfrac{4}{3} - \tfrac{2}{3}\sqrt{7}, 1\right)\\ \tilde{V}_2 &= \left(\tfrac{2}{3}\sqrt{7} - \tfrac{4}{3}, 0 + 1\right)\\ &= \left(-\tfrac{4}{3} + \tfrac{2}{3}\sqrt{7}, 1\right)\end{aligned}$$

and the center

$$\tilde{C} = \left(0 - \tfrac{4}{3}, 0 + 1\right) = \left(-\tfrac{4}{3}, 1\right).$$

1.5.23 Powers and Roots, d'Moivres Theorem

2. We will first transform $-1 - \sqrt{3}\,i$ into polar form. Since the point represented by the complex number $-1 - \sqrt{3}\,i$ is located in the 4th quadrant, we can compute its argument by

$$\begin{aligned}\arg(-1 - \sqrt{3}\,i) &= \arctan\left(\frac{-\sqrt{3}}{-1}\right)\\ &= \arctan\sqrt{3} = \tfrac{\pi}{3}.\end{aligned}$$

Its modulus or absolute value is

$$|-1 - \sqrt{3}\,i| = \sqrt{(-1)^2 + \left(-\sqrt{3}\right)^2} = 2.$$

Therefore, $-1 - \sqrt{3}\,i$ has the polar representation

$$-1 - \sqrt{3}\,i = 2e^{\frac{\pi}{3}i}$$

and by d'Moivres Theorem

$$\begin{aligned}\left(-1 - \sqrt{3}\,i\right)^3 &= \left(2e^{\frac{\pi}{3}i}\right)^3 = 2^3 e^{3\cdot\frac{\pi}{3}i}\\ &= 8e^{\pi i} = 8\left(\cos\pi + i\sin\pi\right)\\ &= 8\left((-1) + 0i\right) = -8.\end{aligned}$$

3. Recall that if $n = 4k + r$ with $0 \le r < 4$ then

$$\begin{aligned}i^n &= i^{4k+r} = i^{4k}i^r = \left(i^4\right)^k i^r = 1^k i^r\\ &= i^r,\end{aligned}$$

and thus

$$i^n = \begin{cases} 1 & n \equiv 0 \pmod 4 \\ i & n \equiv 1 \pmod 4 \\ -1 & n \equiv 2 \pmod 4 \\ -i & n \equiv 3 \pmod 4. \end{cases}$$

In particular, since $39 = 4 \cdot 9 + 3$, we have

$$i^{39} = i^3 = -i.$$

4. We first transform $16 + 12i$ into polar form. Since the point corresponding to the complex number $16 + 12i$ is located in the 1st quadrant, its argument is given by

$$\begin{aligned} \varphi &= \arg(16 + 12i) = \arctan\left(\tfrac{12}{16}\right) \\ &= \arctan\left(\tfrac{3}{4}\right). \end{aligned}$$

Its modulus or absolute value is

$$\begin{aligned} |16 + 12i| &= 4|4 + 3i| = 4\sqrt{4^2 + 3^2} \\ &= 4\sqrt{25} = 20. \end{aligned}$$

Thus

$$16 + 12i = 20e^{\varphi i}.$$

By d'Moivres Theorem, the two square roots of $16 + 12i$ are given by

$$z_0 = \sqrt{20}e^{\frac{\varphi}{2}i} = 2\sqrt{5}e^{\frac{\varphi}{2}i}$$

and

$$z_0 = \sqrt{20}e^{\frac{\varphi}{2}i + \frac{2\pi}{2}} = 2\sqrt{5}e^{\frac{\varphi}{2}i + \pi}.$$

5. Since $64 = 2^6$ is a real number, its argument $\arg(64) = 0$. Thus its polar representation is

$$64 = 2^6 e^{0i}.$$

By d'Moivres Theorem, the six roots of 64 are given by the equations

$$z_k = 2e^{\frac{0i}{6} + \frac{2\pi k i}{6}} = 2\left(\cos\tfrac{\pi k}{3} + i\sin\tfrac{\pi k}{3}\right)$$

for $k = 0, 1, \ldots, 5$

Explicitely

$$\begin{aligned}
z_0 &= 2 \\
z_1 &= 2\big(\cos\tfrac{\pi}{3} + i\sin\tfrac{\pi}{3}\big) \\
&= 2\big(\tfrac{1}{2} + \tfrac{i}{2}\sqrt{3}\big) = 1 + i\sqrt{3} \\
z_3 &= 2\big(\cos\tfrac{2\pi}{3} + i\sin\tfrac{2\pi}{3}\big) \\
&= 2\big(-\tfrac{1}{2} + \tfrac{i}{2}\sqrt{3}\big) = -1 + i\sqrt{3} \\
z_3 &= -2 \\
z_4 &= -1 - i\sqrt{3} \\
z_5 &= 1 - i\sqrt{3}.
\end{aligned}$$

Note that the six roots of 64 are equally spaced around the circle of radius 2 starting with the angle 0 with counterclockwise increments of $\pi/3$.

2

TRIGONOMETRY

2.1 Definitions and Results

Remark 2.1
The Canadarm2 robotic manipulator on the International Space Station is operated by controlling the angles of its joints. Calculating the final position of the astronaut at the end of the arm requires repeated use of the trigonometric functions of those angles.

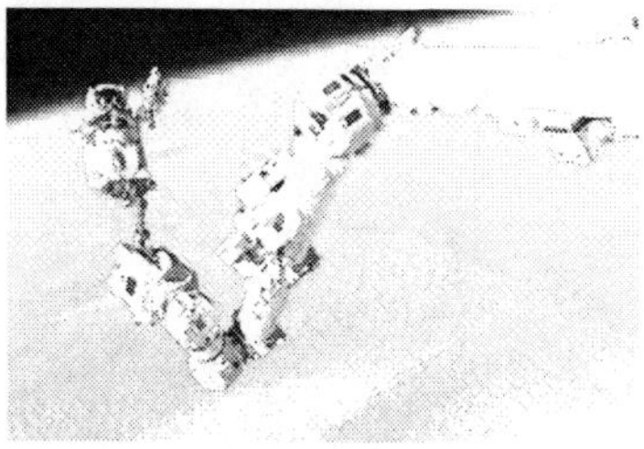

Figure 2.1: *Canadarm2 Robotic Manipulator*

2.1.1 Plane Angles

Definition 2.1 (Plane Angle)
A *plane angle* is generated by rotating a ray about its endpoint, called the vertex. The starting position of the ray is called the initial side of the angle, the final position the terminal side of the angle. An angle is said to be *positive* if it is generated by a counterclockwise rotation, and *negative* if the rotation is clockwise.

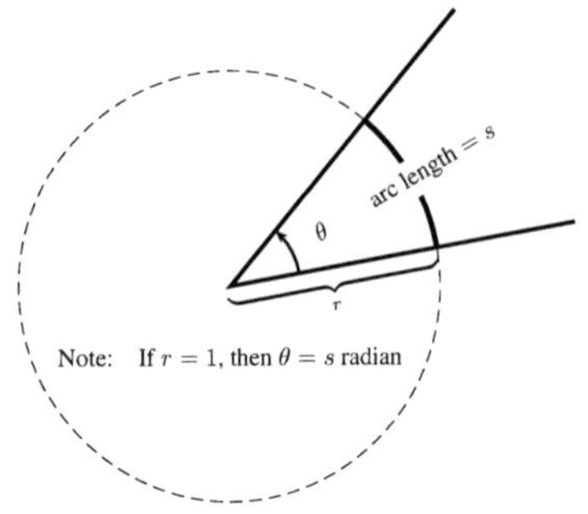

Figure 2.2: *Measuring Angles*

Definition 2.2 (Magnitude or Measure)
The *magnitude or measure* of the angle is the "amount of rotation" generating the angle. To measure an angle θ, consider an arbitrary circular arc centered at the vertex of the angle from the initial side to the terminal side of the angle. The measure θ of the angle is defined as

$$\theta = \frac{sk}{r},$$

where s is the length of the arc, r the radius of the circle and k a scaling constant depending on the angle units used. Note that a full angle has magnitude $2\pi k$. The most common *units* for measuring angles are radian and degree.

1. *Degree:* The *degree* which is denoted by "°" is $1/360$ of a full circle which has measure $360°$. Note that in this case the scaling constant $k = 360$.

2. *Radian:* The *radian*, denoted by "rad" is the angle subtended by an arc of a circle with the same length as its radius. Note that a full angle has measure 2π, hence the scaling constant $k = 1$.

Lemma 2.1 (Conversion: Radian, Degree)
If θ_r and θ_d denote the measure of an angle in radian and degree, respectively, then

$$\frac{\theta_d}{360} = \frac{\theta_r}{2\pi}$$

Example 2.1
Convert 1 radian into degrees!

SOLUTION: Since

$$\theta_d = \frac{\theta_r \cdot 360}{2\pi},$$

we have

$$1\,\text{rad} = \left(\frac{1 \cdot 360}{2\pi}\right)^\circ \approx 57.30^\circ$$

◁

2.1.2 Geometric Definition of Trigonometric Functions

Remark 2.2
Recall that in similar triangles, the ratios of corresponding sides are equal. For the definition of trigonometric functions, we will confine ourselves to right

triangles. If we denote one base angle by θ the other has magnitude $90° - \theta$. Consider the quotients

$$\frac{o}{h} = \frac{o'}{h'}, \quad \frac{a}{h} = \frac{a'}{h'}$$

Clearly, these ratios only depend on the angle θ and constitute invariants of the class of all similar right triangles with base angle θ and are called $\sin(\theta)$, $\cos(\theta)$ etc.

Hence the trigonometric functions for acute angles θ, i.e. $0 < \theta < 90°$ can be defined by

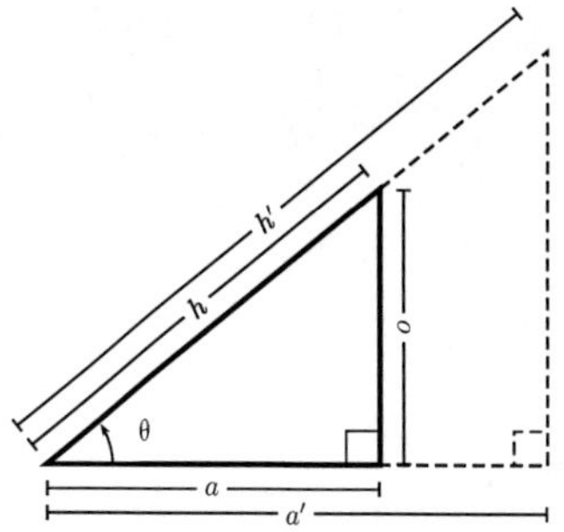

Figure 2.3: *Geometric Definition of the Trigonometric Functions*

Definition 2.3 (Trigonometric Functions)

$$\sin(\theta) = \frac{\text{opposite}}{\text{hypotenuse}} \qquad \cot(\theta) = \frac{\text{adjacent}}{\text{opposite}}$$

$$\cos(\theta) = \frac{\text{adjacent}}{\text{hypotenuse}} \qquad \sec(\theta) = \frac{\text{hypotenuse}}{\text{adjacent}}$$

$$\tan(\theta) = \frac{\text{opposite}}{\text{adjacent}} \qquad \csc(\theta) = \frac{\text{hypotenuse}}{\text{opposite}}$$

Example 2.2 (Important Special Values of Trigonometic Functions)
The following special values of the trigonometric functions can be easily computed using a unit equilateral triangle and unit square, respectively. They should be committed to memory.

θ	Note!	$\sin\theta$	$\cos\theta$	$\tan\theta$	$\cot\theta$
0°	$\frac{1}{2}\sqrt{0}$	0	1	0	∞
30°	$\frac{1}{2}\sqrt{1}$	$\frac{1}{2}$	$\frac{1}{2}\sqrt{3}$	$\frac{1}{3}\sqrt{3}$	$\sqrt{3}$
45°	$\frac{1}{2}\sqrt{2}$	$\frac{1}{2}\sqrt{2}$	$\frac{1}{2}\sqrt{2}$	1	1
60°	$\frac{1}{2}\sqrt{3}$	$\frac{1}{2}\sqrt{3}$	$\frac{1}{2}$	$\sqrt{3}$	$\frac{1}{3}\sqrt{3}$
90°	$\frac{1}{2}\sqrt{4}$	1	0	∞	0

2.1.3 Circular Functions

The trigonometric functions can be defined for all real numbers using the unit circle:

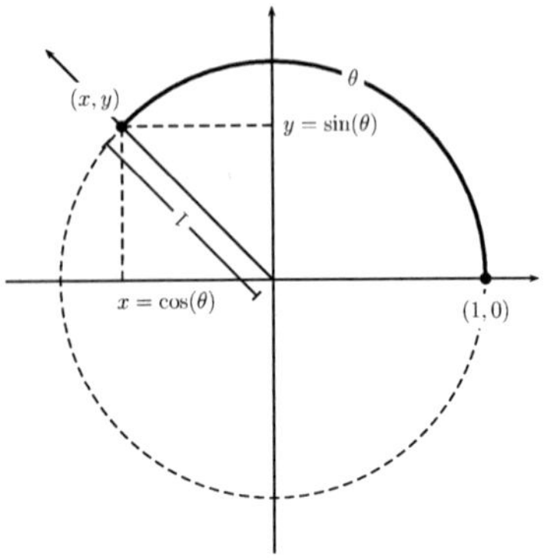

Figure 2.4: *The "Circular Functions"* sin *and* cos

Definition 2.4 (Trigonometric Functions)
For any $\theta \in \mathbb{R}$, and (x, y) denoting the intersection of the terminal side of the angle θ with the unit circle, we define

$$\begin{aligned}
\sin(\theta) &= y \\
\cos(\theta) &= x \\
\tan(\theta) &= \frac{\sin(\theta)}{\cos(\theta)}, &&\text{for } \theta \neq (2n+1)\tfrac{\pi}{2} \\
\cot(\theta) &= \frac{\cos(\theta)}{\sin(\theta)}, &&\text{for } \theta \neq n\pi \\
\sec(\theta) &= \frac{1}{\cos(\theta)}, &&\text{for } \theta \neq (2n+1)\tfrac{\pi}{2} \\
\csc(\theta) &= \frac{1}{\sin(\theta)}, &&\text{for } \theta \neq n\pi
\end{aligned}$$

for all $n \in \mathbb{Z}$.

Definition 2.5 (Periodic)
A function f defined on a set $I \subseteq \mathbb{R}$ is said to be *periodic* if there exists a positive number $p \in \mathbb{R}$ such that

$$f(x + p) = f(x)$$

for all $x \in I$. Any such p is called a *period* of f. The smallest period (if it exists) is called the *fundamental period* or just the *period* of f.

Definition 2.6 (Even, Odd Functions)
Suppose f is defined on the "symmetric" set $I \subseteq \mathbb{R}$, (i.e. $x \in I \Rightarrow -x \in I$). Then f is said to be

1. *even* if $f(-x) = f(x)$ for all $x \in I$,
2. *odd* if $f(-x) = -f(x)$ for all $x \in I$.

Theorem 2.1 (Periodicity and Parity of the Trigonometric Functions)

1. sin *and* cos *have period* 2π.
2. tan *and* cot *have period* π.
3. sec *and* csc *have period* 2π.
4. sin *is odd and* cos *is even*
5. tan *and* cot *are odd*
6. sec *is even,* csc *is odd*

Theorem 2.2 (Zeros)
Let $x \in \mathbb{R}$, then

1. $\sin(x) = 0 \Leftrightarrow x = n\pi, \forall n \in \mathbb{Z}$
2. $\cos(x) = 0 \Leftrightarrow x = (2n+1)\pi/2, \forall n \in \mathbb{Z}$
3. $\tan(x) = 0 \Leftrightarrow x = n\pi, \forall n \in \mathbb{Z}$
4. $\cot(x) = 0 \Leftrightarrow x = (2n+1)\pi/2, \forall n \in \mathbb{Z}$

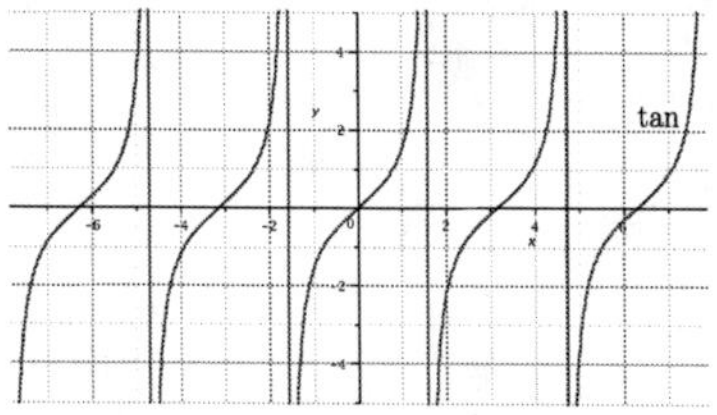

Figure 2.6: tan

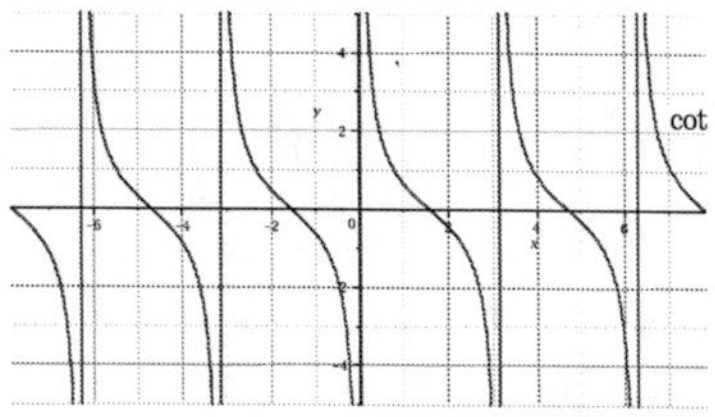

Figure 2.7: cot

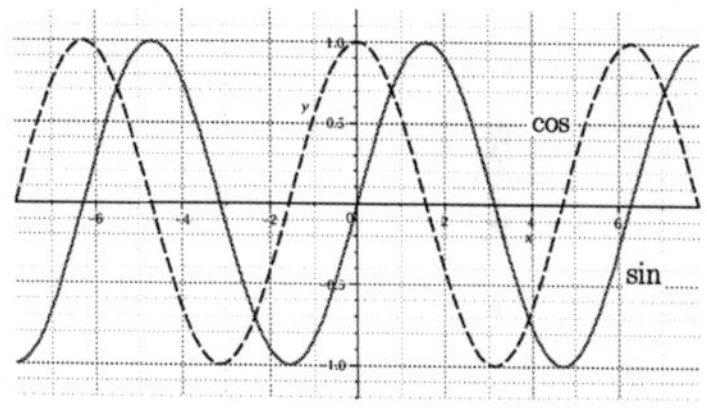

Figure 2.5: sin *and* cos

Example 2.3 (Special Angles of Trigonometric Functions)

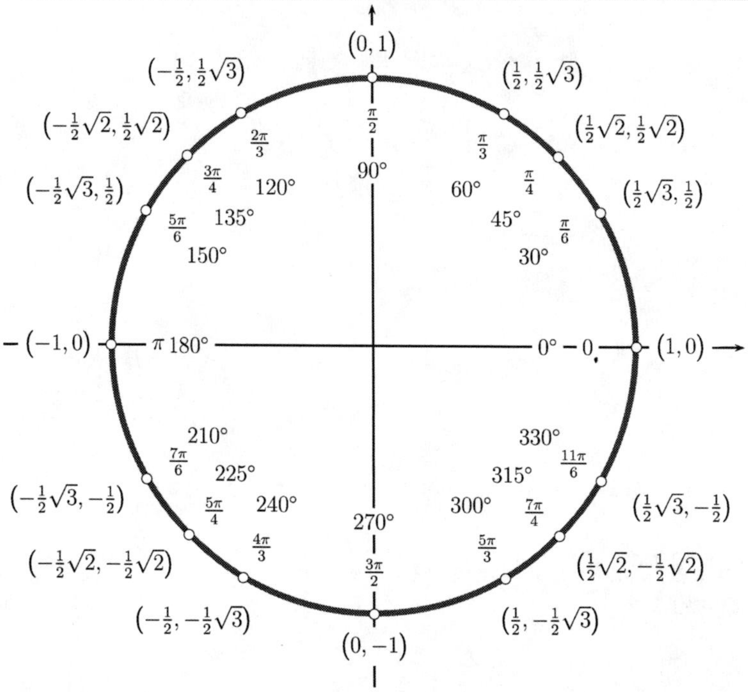

Figure 2.10: *Special Values of* sin *and* cos

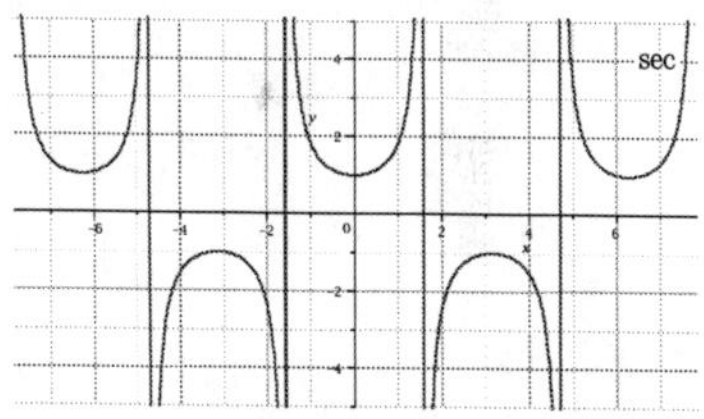

Figure 2.8: sec

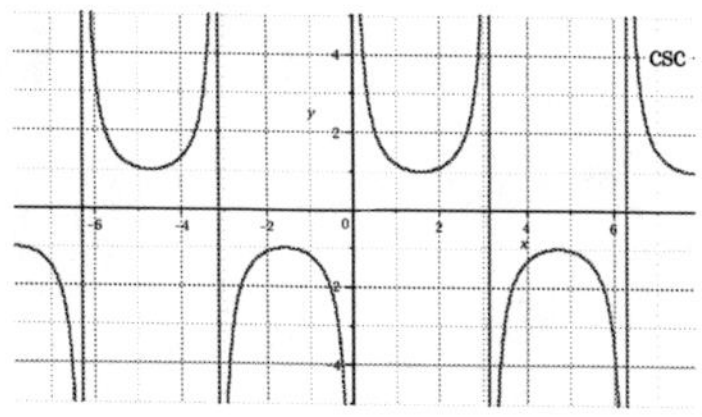

Figure 2.9: csc

2.1.4 Trigonometric Identities

Theorem 2.3 (Pythagorean Identities)
For all $x \in \mathbb{R}$:

1. $\cos^2 x + \sin^2 x = 1$
2. $1 + \tan^2 x = \sec^2 x$
3. $1 + \cot^2 x = \csc^2 x$

provided the expressions are defined.

Theorem 2.4 (Addition Formulae)
For all $x, y \in \mathbb{R}$

1. $\sin(x + y) = \sin x \cos y + \sin y \cos x$
2. $\cos(x + y) = \cos x \cos y - \sin x \sin y$
3. $\tan(x + y) = \dfrac{\tan x + \tan y}{1 - \tan x \tan y}$
4. $\cot(x + y) = \dfrac{\cot x \cot y - 1}{\cot x + \cot y}$

provided the expressions are defined.

Theorem 2.5 (Double-Angle Identities)
For all $x \in \mathbb{R}$

1. $\sin 2x = 2 \sin x \cos x$
2. $\cos 2x = \cos^2 x - \sin^2 x$
3. $\tan 2x = \dfrac{2 \tan x}{1 - \tan^2 x}$
4. $\cot 2x = \dfrac{\cot^2 x - 1}{2 \cot x}$

provided the expressions are defined.

Theorem 2.6 (Power Identities)
For all $x \in \mathbb{R}$

1. $\sin^2 x = \frac{1}{2}(1 - \cos 2x)$
2. $\cos^2 x = \frac{1}{2}(1 + \cos 2x)$

Theorem 2.7 (Product Identities)
For all $x, y \in \mathbb{R}$

1. $\sin x \sin y = \frac{1}{2}\cos(x-y) - \frac{1}{2}\cos(x+y)$
2. $\sin x \cos y = \frac{1}{2}\sin(x-y) + \frac{1}{2}\sin(x+y)$
3. $\cos x \cos y = \frac{1}{2}\cos(x-y) + \frac{1}{2}\cos(x+y)$

Theorem 2.8 (Half-Angle Identities)
For all $x \in \mathbb{R}$

1. $\sin\frac{x}{2} = \pm\sqrt{\frac{1-\cos x}{2}}$
2. $\cos\frac{x}{2} = \pm\sqrt{\frac{1+\cos x}{2}}$
3. $\tan\frac{x}{2} = \frac{\sin x}{1+\cos x} = \frac{1-\cos x}{\sin x}$

provided the expressions are defined.

Theorem 2.9 (Phase Shift)
For all $x \in \mathbb{R}$

1. $\sin\left(\frac{\pi}{2}-x\right) = \cos x$
2. $\cos\left(\frac{\pi}{2}-x\right) = \sin x$
3. $\tan\left(\frac{\pi}{2}-x\right) = \cot x$

Figure 2.11: *Leonhard Euler (1707–1783)*

Theorem 2.10 (Euler Formula)
For all $x \in \mathbb{R}$

$$e^{ix} = \cos x + i\sin x$$

Corollary 2.1
For all $x \in \mathbb{R}$

1. $\cos x = \frac{1}{2}\left(e^{ix} + e^{-ix}\right)$
2. $\sin x = \frac{1}{2i}\left(e^{ix} - e^{-ix}\right)$

Theorem 2.11 (Derivatives)
For all $x \in \mathbb{R}$

1. $\sin'(x) = \cos(x)$
2. $\cos'(x) = -\sin(x)$
3. $\tan'(x) = \sec^2(x) = 1 + \tan^2(x)$
4. $\cot'(x) = -\csc^2(x) = -1 - \cot^2(x)$
5. $\sec'(x) = \sec(x)\tan(x)$
6. $\csc'(x) = -\csc(x)\cot(x)$

provided the expressions are defined.

Theorem 2.12 (Derivatives of the Inverse Trigonometric Functions)
For all $x \in \mathbb{R}$

1. $(\sin^{-1})'(x) = \dfrac{1}{\sqrt{1-x^2}}, \quad |x| < 1$
2. $(\cos^{-1})'(x) = \dfrac{-1}{\sqrt{1-x^2}}, \quad |x| < 1$
3. $(\tan^{-1})'(x) = \dfrac{1}{1+x^2}$
4. $(\cot^{-1})'(x) = \dfrac{-1}{1+x^2}$
5. $(\sec^{-1})'(x) = \dfrac{1}{|x|\sqrt{x^2-1}}, \quad |x| > 1$
6. $(\csc^{-1})'(x) = \dfrac{-1}{|x|\sqrt{x^2-1}}, \quad |x| > 1$

2.1.5 Taylor Series Expansions of Trigonometric Functions

Figure 2.12: *Jakob Bernoulli (1654–1705)*

Theorem 2.13 (Taylor Series Expansions of $\sin$ **and** $\cos$**)**
For all $x \in \mathbb{R}$

1. $\sin(x) = \sum_{n=0}^{\infty} (-1)^n \frac{x^{2n+1}}{(2n+1)!}$

2. $\cos(x) = \sum_{n=0}^{\infty} (-1)^n \frac{x^{2n}}{(2n)!}$

3. $\tan(x) = \sum_{n=0}^{\infty} (-1)^{n-1} \frac{2^{2n}(2^{2n}-1)B_{2n}\, x^{2n-1}}{(2n)!}, \quad |x| < \frac{\pi}{2}$

4. $\cot(x) = \sum_{n=0}^{\infty} (-1)^n \frac{2^{2n} B_{2n}\, x^{2n-1}}{(2n)!}, \quad 0 < |x| < \pi$

5. $\sec(x) = \sum_{n=0}^{\infty} (-1)^n \frac{E_{2n}\, x^{2n}}{(2n)!}, \quad |x| < \frac{\pi}{2}$

6. $\csc(x) = \sum_{n=0}^{\infty} (-1)^{n+1} \frac{2(2^{2n-1}-1)B_{2n}\, x^{2n-1}}{(2n)!}, \quad 0 < |x| < \pi$

Remark 2.3
Note that B_n and E_n denote the n-th Bernoulli and Euler number, respectively.

The *Bernoulli numbers,* named after Jakob Bernoulli, can be defined by the generating function

$$\frac{t}{e^t - 1} = \sum_{n=0}^{\infty} B_n \frac{t^n}{n!}$$

The $B - n$ are all rational numbers, $B_{2n+1} = 0$ for $n \geq 1$ and the (non-zero) Bernoulli numbers alternate in sign. The Bernoulli numbers can also be computed using the recursion

$$B_0 = 1, \qquad \sum_{k=0}^{n-1} \binom{n}{k} B_k \quad \text{for } n = 2, 3, \ldots$$

Thus $B_0 = 1$, $B_1 = -\frac{1}{2}$, $B_2 = \frac{1}{6}$, $B_3 = 0$, $B_4 = -\frac{1}{3}$, $B_5 = 0$, $B_6 = \frac{1}{42}$, $B_7 = 0$, $B_8 = -\frac{1}{3}$, $B_9 = 0$, $B_{10} = \frac{5}{66}, \ldots$

Figure 2.13: *Leonhard Euler (1707–1783)*

The *Euler numbers* or *secant numbers*, named after Leonhard Euler, can be defined by the generating function

$$\frac{2e^{t/2}}{e^t + 1} = \sum_{n=0}^{\infty} \frac{E_n \cdot t^n}{2^n \cdot n!} = \sum_{n=0}^{\infty} \frac{E_n}{n!} \left(\frac{t}{2}\right)^n$$

Some value of the Euler numbers are: $E_0 = 1$, $E_1 = 0$, $E_2 = -1$, $E_3 = 0$, $E_4 = 5$, $E_5 = 0$, $E_6 = -61$, $E_7 = 0$, $E_8 = 1385$, $E_9 = 0$, $E_{10} = -50521$, ...

2.1.6 The Laws of Sines and Cosines

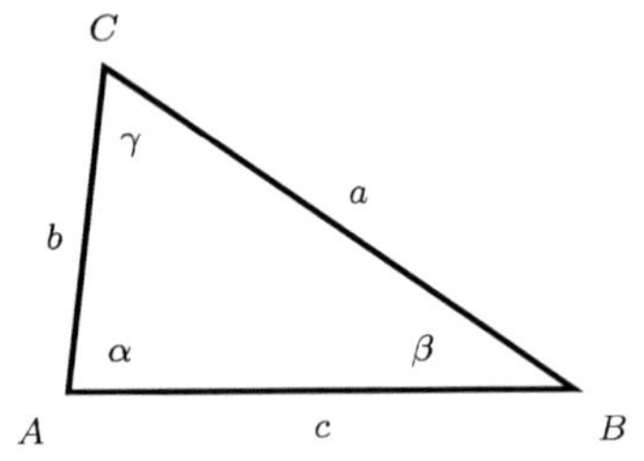

Figure 2.14:

Theorem 2.14 (Law of Sines)
In any triangle ΔABC, the length of the sides are proportional to the sines of the opposite angles; i.e.

$$\frac{a}{\sin(\alpha)} = \frac{b}{\sin(\beta)} = \frac{c}{\sin(\gamma)}$$

where the sides and angles are labeled as usual (see the figure in the margin).

Theorem 2.15 (Law of Cosines)
In any triangle ΔABC, we have

$$\begin{aligned} a^2 &= b^2 + c^2 - 2bc\cos\alpha \\ b^2 &= a^2 + c^2 - 2ac\cos\beta \\ c^2 &= a^2 + b^2 - 2ab\cos\gamma \end{aligned}$$

Example 2.4 (Area of a Triangle I)
Given a triangle ΔABC, where the sides and angles are denoted as in the figure above. Then its area F expressed in terms of the lengths of two sides and the enclosed angle is

$$F = \tfrac{1}{2}\, bc\sin(\alpha) = \tfrac{1}{2}\, ab\sin(\gamma) = \tfrac{1}{2}\, ac\sin(\beta)$$

Example 2.5 (Area of a Triangle II)
Using the "Law of Cosine", we can express the area of a triangle in terms of the length of its three sides as

$$\begin{aligned} F &= \frac{1}{4}\sqrt{4b^2c^2 - \left(b^2 + c^2 - a^2\right)^2} \\ &= \frac{1}{4}\sqrt{(a+b-c)(a-b+c)(a+b+c)(-a+b+c)} \end{aligned}$$

Example 2.6 (Area of a Triangle III, Heron's Formula)
Let $s := \frac{1}{2}(a+b+c)$ denote the semi-perimeter of the triangle ΔABC, then the previous formula can be simplified to

$$F = \sqrt{s(s-a)(s-b)(s-c)}$$

2.1.7 Inverse Trigonometric Functions

The trigonometric functions are multivalued and therefore not invertible. However, by restricting them to their "principal branch", they can be made one-to-one or injective and therefore invertible:

Theorem 2.16 (Inverse Trigonometric Functions)

1. Arcsine: *The inverse of the restriction of* sin *to its principal domain* $[-\frac{\pi}{2}, \frac{\pi}{2}]$ *is the function* $\arcsin : [-1, 1] \to [-\frac{\pi}{2}, \frac{pi}{2}]$ *which is implicitly defined by*
$$\arcsin(x) = y \quad \Leftrightarrow \quad \sin(y) = x$$

2. Arccosine: *The inverse of the restriction of* cos *to its principal domain* $[0, \pi]$ *is the function* $\arccos : [-1, 1] \to [0, \pi]$ *which is implicitly defined by*
$$\arccos(x) = y \quad \Leftrightarrow \quad \cos(y) = x$$

3. Arctangent: *The inverse of the restriction of* tan *to its principal domain* $(-\frac{\pi}{2}, \frac{\pi}{2})$ *is the function* $\arctan : \mathbb{R} \to (-\frac{\pi}{2}, \frac{\pi}{2})$ *which is implicitly defined by*
$$\arctan(x) = y \quad \Leftrightarrow \quad \tan(y) = x$$

4. Arccotangent: *The inverse of the restriction of* cot *to its principal domain* $(0, \pi)$ *is the function* $\operatorname{arccot} : \mathbb{R} \to (0, \pi)$ *which is implicitly defined by*
$$\operatorname{arccot}(x) = y \quad \Leftrightarrow \quad \cot(y) = x$$

5. Arcsecant: *The inverse of the restriction of* sec *to its principal domain* $[0, \pi] \setminus \{\frac{\pi}{2}\}$ *is the function* $\operatorname{arcsec} : \mathbb{R} \setminus (-1, 1) \to [0, \pi] \setminus \{\frac{\pi}{2}\}$ *which is implicitly defined by*
$$\operatorname{arcsec}(x) = y \quad \Leftrightarrow \quad \sec(y) = x$$

6. Arccosecant: *The inverse of the restriction of* csc *to its principal domain* $[\frac{\pi}{2}, \frac{\pi}{2}] \setminus \{0\}$ *is the function* $\operatorname{arccsc} : \mathbb{R} \setminus (-1, 1) \to [-\frac{\pi}{2}, \frac{\pi}{2}] \setminus \{0\}$ *which is implicitly defined by*
$$\operatorname{arccsc}(x) = y \quad \Leftrightarrow \quad \csc(y) = x$$

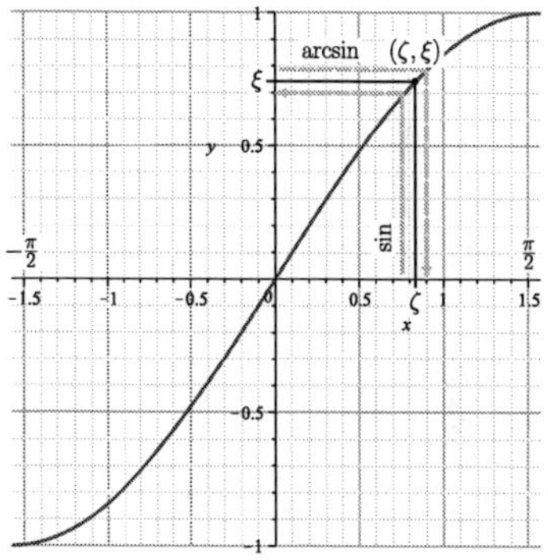

Figure 2.15: *The Restriction of Sine to Its Principal Domain*
$\sin \zeta = \xi \Leftrightarrow \arcsin \xi = \zeta$

Remark 2.4
The inverse trigonometric functions are often denoted by $\sin^{-1}$, $\cos^{-1}$, etc. However this notation where the exponent -1 refers to composition "$\circ$" easily creates a conflict with the common semantics for expressions like $\sin^2(x)$ where the exponents refer to ordinary multiplication.

2.1.8 Hyperbolic Functions

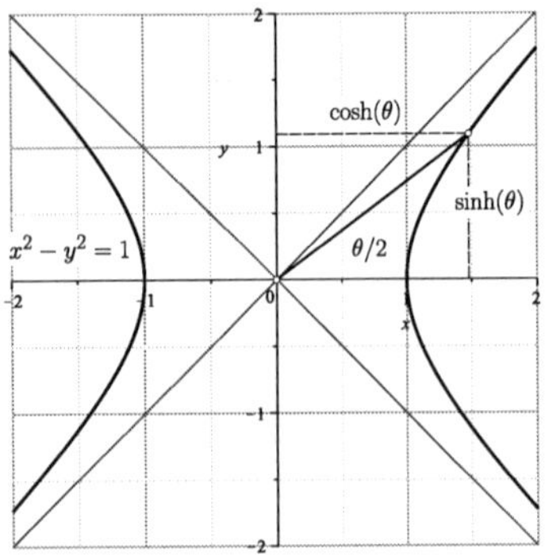

Figure 2.16: *Geometric Definition of* sinh *and* cosh

The ray through the origin intercepts the hyperbola $x^2 - y^2 = 1$ in the point $(\cosh\theta, \sinh\theta)$, where θ equals twice the area between the ray, the right branch of the hyperbola and the x-axis.

Remark 2.5
The hyperbolic functions were introduced in the 18th century by the Swiss mathematician Johann Heinrich Lambert. They are analogs to the trigonometric or "circular functions". As the points

$$C := \{(\cos\theta, \sin\theta) \mid \theta \in \mathbb{R}\}$$

form the "unit circle" $x^2 + y^2 = 1$, so the points

$$H := \{(\cosh\theta, \sinh\theta) \mid \theta \in \mathbb{R}\}$$

form the right branch of the "unit hyperbola" $x^2 + y^2 = 1$.

Hyperbolic functions occur naturally in the solutions of some important linear differential equations.

Definition 2.7 (Hyperbolic Functions)
For all $x \in \mathbb{R}$, we define

$$\begin{aligned}
\sinh x &= \tfrac{1}{2}\left(e^x - e^{-x}\right)\\
\cosh x &= \tfrac{1}{2}\left(e^x + e^{-x}\right)\\
\tanh x &= \frac{\sinh x}{\cosh x} = \frac{e^{2x} - 1}{e^{2x} + 1}\\
\coth x &= \frac{\cosh x}{\sinh x} = \frac{e^{2x} + 1}{e^{2x} - 1}, \quad x \neq 0\\
\operatorname{sech} x &= \frac{1}{\cosh x} = \frac{2}{e^x + e^{-x}}\\
\operatorname{csch} x &= \frac{1}{\sinh x} = \frac{2}{e^x - e^{-x}}, \quad x \neq 0
\end{aligned}$$

Figure 2.17:

Theorem 2.17 (Parity of the Hyperbolic Functions)

1. sinh *is odd and* cosh *is even.*
2. tanh *and* coth *are odd.*
3. sech *is even and* csch *is odd.*

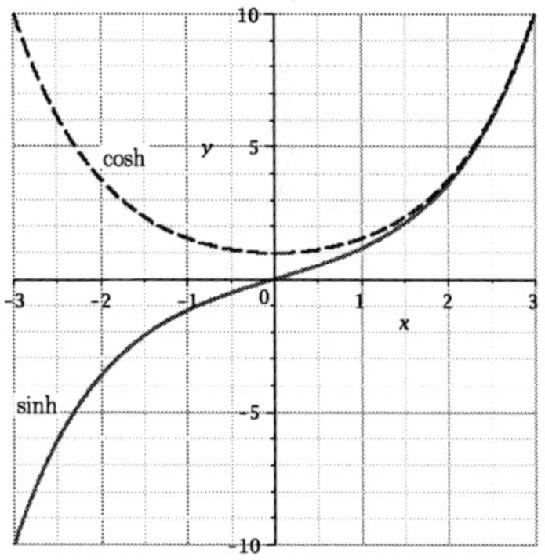

Figure 2.18: sinh *and* cosh

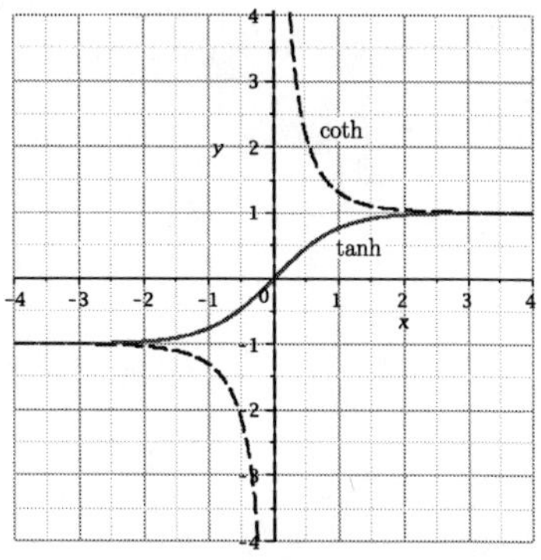

Figure 2.19: tanh *and* coth

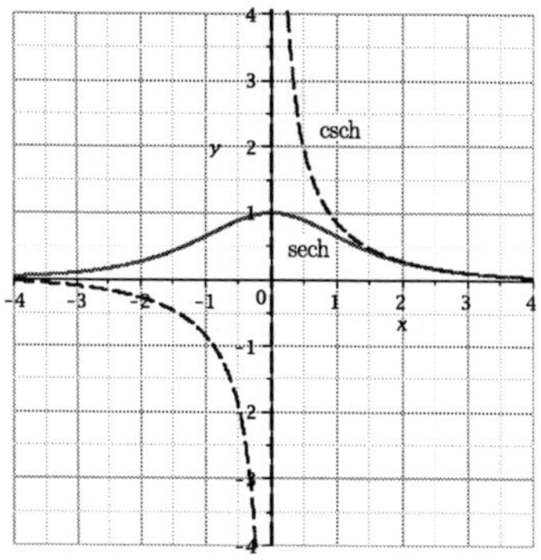

Figure 2.20: *sech and csch*

2.1.9 Hyperbolic Identities

Theorem 2.18 (Pythagorean Identities)
For all $x \in \mathbb{R}$

1. $\cosh^2 x - \sinh^2 = 1$
2. $\tanh^2 x + \operatorname{sech}^2 x = 1$

Theorem 2.19 (Addition Formulae)
For all $x, y \in \mathbb{R}$

1. $\sinh(x+y) = \sinh x \cosh y + \sinh y \cosh x$
2. $\cosh(x+y) = \cosh x \cosh y + \sinh x \sinh y$
3. $\tanh(x+y) = \dfrac{\tanh x + \tanh y}{1 + \tanh x \tanh y}$

Theorem 2.20 (Double-Angle Identities)
For all $x \in \mathbb{R}$

1. $\sinh 2x = 2 \sinh x \cosh x$
2. $\cosh 2x = \cosh^2 x + \sinh^2 x = 2\cosh^2 x - 1 = 2\sinh^2 x + 1$
3. $\tanh 2x = \dfrac{2 \tanh x}{1 + \tanh^2 x}$

Theorem 2.21 (Power Identities)
For all $x \in \mathbb{R}$

1. $\sinh^2 x = \frac{1}{2}(\cosh 2x - 1)$
2. $\cosh^2 x = \frac{1}{2}(\cosh 2x + 1)$

Theorem 2.22 (Half-Angle Identities)
For all $x \in \mathbb{R}$

1. $\sinh \frac{x}{2} = \operatorname{sgn}(x)\sqrt{\dfrac{\cosh x - 1}{2}}$
2. $\cosh \frac{x}{2} = \sqrt{\dfrac{\cosh x + 1}{2}}$
3. $\tanh \frac{x}{2} = \dfrac{\sinh x}{\cosh x + 1} = \dfrac{\cosh x - 1}{\sinh x}, \quad x \neq 0$

Theorem 2.23 (Euler Type Formula)
For all $x \in \mathbb{R}$

1. $e^x = \cosh x + \sinh x$
2. $e^{-x} = \cosh x - \sinh x$

2.1.10 Derivatives of Hyperbolic Functions

Theorem 2.24 (Derivatives)
For all $x \in \mathbb{R}$

1. $\sinh'(x) = \cosh(x)$
2. $\cosh'(x) = \sinh(x)$
3. $\tanh'(x) = \operatorname{sech}^2(x) = 1 - \tanh^2(x)$
4. $\coth'(x) = -\cosh^2(x) = 1 - \coth^2(x), \quad x \neq 0$
5. $\operatorname{sech}'(x) = -\operatorname{sech}(x)\tanh(x)$
6. $\operatorname{csch}'(x) = -\operatorname{csch}(x)\coth(x), \quad x \neq 0$

Theorem 2.25 (Derivatives of Inverse Hyperbolic Functions)
For all $x \in \mathbb{R}$

1. $(\sinh^{-1})'(x) = \dfrac{1}{\sqrt{x^2+1}}$
2. $(\cosh^{-1})'(x) = \dfrac{1}{\sqrt{x^2-1}} \quad |x| > 1$
3. $(\tanh^{-1})'(x) = \dfrac{1}{1-x^2} \quad |x| \neq 1$
4. $(\coth^{-1})'(x) = \dfrac{1}{1-x^2} \quad |x| \neq 1$
5. $(\text{sech}^{-1})'(x) = \dfrac{-1}{x\sqrt{1-x^2}}, \quad |x| < 1$
6. $(\text{csch}^{-1})'(x) = \dfrac{-1}{|x|\sqrt{1+x^2}}$

2.1.11 Integrals of Hyperbolic Functions

Theorem 2.26 (Integrals)
For all $x \in \mathbb{R}$

1. $\displaystyle\int \sinh x \, dx = \cosh x + C$
2. $\displaystyle\int \cosh x \, dx = \sinh x + C$
3. $\displaystyle\int \tanh x \, dx = \ln(\cosh x) + C$
4. $\displaystyle\int \coth x \, dx = \ln|\sinh x| + C, \quad x \neq 0$

Theorem 2.27 (Integrals of Inverse Hyperbolic Functions)
For all $x \in \mathbb{R}$

1. $\displaystyle\int \frac{dx}{\sqrt{x^2+1}} = \sinh^{-1}(x) + C$
2. $\displaystyle\int \frac{dx}{\sqrt{x^2-1}} = \cosh^{-1}(x) + C, \quad |x| \geq 1$
3. $\displaystyle\int \frac{dx}{1-x^2} = \tanh^{-1}(x) + C, \quad |x| < 1$
4. $\displaystyle\int \frac{dx}{1-x^2} = \coth^{-1}(x) + C, \quad |x| > 1$
5. $\displaystyle\int \frac{dx}{x\sqrt{1+x^2}} = -\operatorname{csch}^{-1}(x) + C$
6. $\displaystyle\int \frac{dx}{x\sqrt{1-x^2}} = -\operatorname{sech}^{-1}(x) + C, \quad |x| < 1$

2.1.12 Taylor/Laurent Series Expansions of Hyperbolic Functions

Theorem 2.28 (Taylor/Laurent Series Expansions)
For all $x \in \mathbb{R}$

1. $\displaystyle\sinh(x) = \sum_{n=0}^{\infty} \frac{x^{2n+1}}{(2n+1)!}$
2. $\displaystyle\cosh(x) = \sum_{n=0}^{\infty} \frac{x^{2n}}{(2n)!}$
3. $\displaystyle\tanh(x) = \sum_{n=1}^{\infty} \frac{2^{2n}(2^{2n}-1)B_{2n}\,x^{2n-1}}{(2n)!}, \quad |x| < \frac{\pi}{2}$
4. $\displaystyle\coth(x) = \frac{1}{x} + \sum_{n=1}^{\infty} \frac{2^{2n}B_{2n}\,x^{2n-1}}{(2n)!}, \quad 0 < |x| < \pi$
5. $\displaystyle\operatorname{sech}(x) = \sum_{n=0}^{\infty} \frac{E_{2n}\,x^{2n}}{(2n)!}, \quad |x| < \frac{\pi}{2}$
6. $\displaystyle\operatorname{csch}(x) = \frac{1}{x} + \sum_{n=1}^{\infty} \frac{2(1-2^{2n-1})B_{2n}\,x^{2n-1}}{(2n)!}, \quad 0 < |x| < \pi$

Remark 2.6
Note that B_n and E_n denote the n-th Bernoulli and Euler numbers, respectively. See Remark 2.3 on page 109.

2.1.13 Inverse Hyperbolic Functions

The inverse hyperbolic functions can be expressed as logarithms:

Theorem 2.29 (Inverse Hyperbolic Functions as Logarithms)
For all $x \in \mathbb{R}$

1. $\sinh^{-1}(x) = \ln\big(x + \sqrt{x^2+1}\big)$
2. $\cosh^{-1}(x) = \ln\big(x + \sqrt{x^2-1}\big), \quad x \geq 1$
3. $\tanh^{-1}(x) = \frac{1}{2}\ln\left(\dfrac{1+x}{1-x}\right), \quad |x| < 1$
4. $\coth^{-1}(x) = \frac{1}{2}\ln\left(\dfrac{x+1}{x-1}\right), \quad |x| > 1$
5. $\operatorname{sech}^{-1}(x) = \ln\left(\dfrac{1+\sqrt{1-x^2}}{x}\right), \quad 0 < x \leq 1$
6. $\operatorname{csch}^{-1}(x) = \ln\left(\dfrac{1}{x} + \dfrac{\sqrt{1+x^2}}{|x|}\right)$

2.2 Exercises

2.2.1 Simplifying Trigonometric Expressions

1. $\sin x \cos x + \cos^2 x$ has

2. $2 \sin x \cos^3 x - 2 \sin^3 x \cos x$ has

3. $\dfrac{\sec^4 x - \tan^4 x}{\sec^2 x + \tan^2 x}$ has

4. $\dfrac{2 \sin^2 x}{\cos^3 x} \cdot \left(\dfrac{\cos x}{2 \sin x}\right)^2$ has

5. $\dfrac{3 \sin x}{\cos^2 x} \cdot \dfrac{\cos^2 x + \cos x \sin x}{\sin^2 x - \cos^2 x}$ has

6. $\dfrac{2 \tan \theta}{1 + \tan^2 \theta}$ has

7. $\dfrac{\tan x - \sin x}{2 \tan x}$ has

8. $\sin(\alpha + \beta) \sin(\alpha - \beta)$ has

9. $\left(\dfrac{\cot x}{\csc x}\right)^2 + \dfrac{1}{\csc^2 x}$ has

10. $\dfrac{\cos \theta + \sin \theta}{\cos \theta}$ has

11. $(\cos^2 x) \dfrac{1 + \tan^2 x}{1 - \cos^2 x}$ has

2.2.2 Sum and Difference Formulas for Sine and Cosine

1. Give a formula for $\sin(x+y)$ in terms of $\cos x$, $\sin x$, $\sin y$ and $\cos y$. has

2. Give a formula for $\cos(x+y)$ in terms of $\cos x$, $\sin x$, $\sin y$ and $\cos y$. has

3. Give a formula for $\sin(x-y)$ in terms of $\cos x$, $\sin x$, $\sin y$ and $\cos y$. has

4. Give a formula for $\cos(x-y)$ in terms of $\cos x$, $\sin x$, $\sin y$ and $\cos y$. has

2.2.3 Pythagorean Identities

1. Recall the Three Pythagorean Identities has
2. Derive the First Pythagorean identity has
3. Derive the Second Pythagorean identity. has

2.2.4 Double and Half Angle Formulas

1. Express $\sin 2x$ in terms of $\sin x$ and $\cos x$ has
2. Express $\cos 2x$ in terms of $\sin x$ and $\cos x$ has
3. Express $\tan 2x$ in terms of $\tan x$ has
4. Give a formula for $\cos^2 x$ in terms of $\cos 2x$ has
5. Give a formula for $\sin^2 x$ in terms of $\cos 2x$ has
6. Give a formula for $\sin \frac{x}{2}$ in terms of $\cos x$ has
7. Give a formula for $\cos \frac{x}{2}$ in terms of $\cos x$ has
8. Give a formula for $\tan \frac{x}{2}$ in terms of $\cos x$ has

2.2.5 The Laws of Sines and Cosines

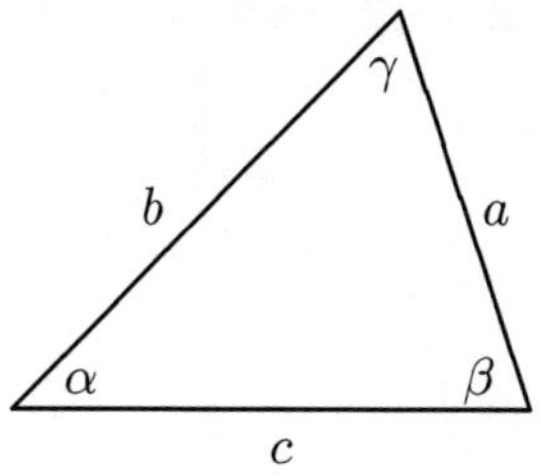

1. $c = 5$, $b = 3$ and $\alpha = \frac{2}{3}\pi$. has
2. $a = 15$, $c = 7$, $\beta = 62°$. has
3. $a = 3.6$, $b = 6.2$, $c = 4.1$ has
4. $\alpha = 36.5°$, $a = 24$, $b = 34$ has
5. $\alpha = \frac{1}{4}\pi$, $\gamma = \frac{1}{3}\pi$, $a = 4$ has

2.2.6 Special Values of Trigonometric Functions

1. $\sin \pi$ a
2. $\cos\left(-\frac{\pi}{3}\right)$ a
3. $\tan\left(\frac{11\pi}{4}\right)$ a
4. $\sin\left(\frac{5\pi}{4}\right)$ a
5. $\csc\left(\frac{3\pi}{4}\right)$ a
6. $\tan\left(\frac{\pi}{6}\right)$ a
7. $\sin^9\left(\frac{\pi}{6}\right)\cos(4\pi)$ a

2.2.7 Simplify Trigonometric Expression Involving Inverse Trigonometric Functions

1. $\sin^{-1}\left(\sin\left(\frac{7\pi}{4}\right)\right)$ a
2. $\sin\left(\cos^{-1}\left(-\frac{1}{2}\right)\right)$ a
3. $\cos\left(\sec^{-1}(2)\right)$ a
4. $\cos\left(\tan^{-1}(x)\right)$ a
5. $\csc\left(\cot^{-1}\left(\frac{x}{2}\right)\right)$ a
6. $\sin\left(\cos^{-1}(x)\right)$ a
7. $\sec\left(\sin^{-1}(x)\right)$ a

2.2.8 Graphing Trigonometric Functions

1. $y = \tan x$ as
2. $y = \cot x$ as
3. $y = \sin x$ as
4. $y = \cos x$ as
5. $y = \sec x$ as
6. $y = \csc x$ as
7. $y = 2\sin\left(\frac{1}{2}x\right)$ as
8. $y = -\cos\left(x - \frac{\pi}{2}\right)$ as

9. $y = \sin(x - \pi) - 2$ as

10. $y = \frac{1}{2}\cos\left(2x + \frac{\pi}{2}\right)$ as

2.2.9 Trigonometric Equations

1. $\sin 2x = 1$ has

2. $\sin x = \frac{1}{2}$ has

3. $\tan\left(\frac{1}{2}x\right) = 1$ has

4. $\sqrt{\sin^2 x} = 1$ has

5. $\sin 2x = -\frac{1}{2}\sqrt{3}$ has

6. $\cos x = \frac{1}{2}\sqrt{2}$ ha

7. $\cos^2 x + 6\cos x + 4 = 0$ has

8. $\sin 2x \cos x + \sin x = 0$ has

9. $\sec^2 x + 2\tan x = 6$ has

2.2.10 Converting Cartesian into Polar Coordinates

1. $(0, 1)$ a
2. $(-3, 0)$ a
3. $(2, -2)$ a
4. $\left(\sqrt{x}, \sqrt{x}\right)$ a
5. $\left(3, -3\sqrt{3}\right)$ a
6. $\left(\sqrt{3}, -1\right)$ a

2.2.11 Converting Polar into Cartesian Coordinates

1. $(1, \frac{1}{3}\pi)$ a
2. $(3, \frac{3}{4}\pi)$ a
3. $(-2, 0)$ a
4. $(-3, -\frac{1}{3}\pi)$ a
5. $(3, -\frac{1}{2}\pi)$ a
6. $(2, 3\pi)$ a
7. $(-1, \frac{1}{4}\pi)$ a
8. $(\frac{1}{3}, \frac{2}{3}\pi)$ a

2.2.12 Hyperbolic Functions

1. $\sinh x$ a
2. $\cosh x$ a
3. $\tanh x$ a
4. $\coth x$ a
5. $\operatorname{sech} x$ a
6. $\operatorname{csch} x$ a

2.2.13 Hyperbolic Identies

1. $\cosh^2 x - \sinh^2 x$ a
2. $\sinh x \cosh y + \sinh y \cosh x$ a
3. $\cosh x \cosh y + \sinh x \sinh y$ a
4. $\sinh x \cosh x$ a
5. $\cosh^2 x + \sinh^2 x$ a
6. $\dfrac{\cosh^2 x - \sinh^2 x}{\tanh^2 x}$ a

2.3 Hints to the Exercises

2.3.1 Simplifying Trigonometric Expressions

1. Use the Double-Angle Identity

$$\sin(2x) = 2\sin x \cos x$$

and the Power Identity

$$\cos^2 x = \tfrac{1}{2}(1 + \cos 2x)$$

2. Factor out $2\sin x \cos x$ and use the Double-Angle Identities.
3. Factor the numerator using $a^2 - b^2 = (a-b)(a+b)$ and apply the Pythagorean Identity $1 + \tan^2 x = \sec^2 x$.
4. Simplify the term.
5. Reduce the fraction! First factor out $\sin x \cos x$ in the numerator.
6. Apply the Pythagorean Identity $1 + \tan^2\theta = \sec^2\theta$, rewrite the fraction in terms of sin and cos and reduce it!
7. Separate terms and rewrite in terms of sin and cos.
8. First use the Addition formulae for sin and expand the expression. Then apply the Power Identities and expand again.
9. Combine the fractions and apply the third Pythagorean Identity.
10. Separate the fraction.
11. Apply the suitable Pythagorean identities to the fraction.

2.3.2 Sum and Difference Formulas for Sine and Cosine

1. See Summary Sheet!
2. See Summary Sheet!
3. See Summary Sheet!
4. See Summary Sheet!

2.3.3 Pythagorean Identities

1. See Summary Sheet!
2. Look at the definition of sin and cos using the unit circle
3. Express tan and sec in terms of sin and cos and apply the First Pythagorean identity!

2.3.4 Double and Half Angle Formulas

1. See Summary Sheet!
2. See Summary Sheet!
3. See Summary Sheet!
4. See Summary Sheet!
5. See Summary Sheet!
6. See Summary Sheet!
7. See Summary Sheet!
8. See Summary Sheet!

2.3.5 The Laws of Sines and Cosines

1. To compute a use the law of cosine, β can be computed with the law of sine.
2. The side b can be computed using the law of cosine, the angle α with the law of sine.
3. Use the Law of Cosines to compute two angles!
4. First compute β using the Law of Sines, then γ, then apply the Law of Cosines to compute c.
5. First compute β, then use the Law of Sines to compute c and b!

2.3.6 Special Values of Trigonometric Functions

2.3.7 Simplify Trigonometric Expression Involving Inverse Trigonometric Functions

2.3.8 Graphing Trigonometric Functions

2.3.9 Trigonometric Equations

1. Consult the definition of sin using the "unit circle"!
2. Consult the definition of sin using the "unit circle"
3. Visulize the definition of $\tan$ using the unit circle

4. Square both sides of the equation. Recall that $\sqrt{A^2} = |A|$.

5. Consult the definition of sin using the unit circle.

6. Consult the definition of cos using the unit circle.

7. Use the substitution $x = \cos x$ to reduce the trigonometric equation to a quadratic equation.

8. Rewrite the equation using the Double-Angle Identity $\sin 2x = \sin x \cos x$ and then factor it.

9. Use the Pythagorean Identity $1 + \tan^2 x = \sec^2 x$ and the substitution $y = \tan x$.

2.3.10 Converting Cartesian into Polar Coordinates

2.3.11 Converting Polar into Cartesian Coordinates

2.3.12 Hyperbolic Functions

2.3.13 Hyperbolic Identies

2.4 Answers to the Exercises

2.4.1 Simplifying Trigonometric Expressions

1. $\frac{1}{2}\big(1+\cos(2x)+\sin(2x)\big)$
2. $\frac{1}{2}\sin(4x)$
3. 1
4. $\frac{1}{2}\sec(x)$
5. $\dfrac{3\tan(x)}{\sin x-\cos x}$
6. $\sin(2\theta)$
7. $\frac{1}{2}\big(1-\cos x\big)$
8. $\frac{1}{2}\cos(2\beta)-\frac{1}{2}\cos(2\alpha)]$
9. 1
10. $1+\tan\theta$
11. $\csc^2 x$

2.4.2 Sum and Difference Formulas for Sine and Cosine

1. $\sin(x+y)=\sin x\cos y+\sin y\cos x$
2. $\cos(x+y)=\cos x\cos y-\sin x\sin y$
3. $\sin(x-y)=\sin x\cos y-\sin y\cos x$
4. $\cos(x-y)=\cos x\cos y+\sin x\sin y$

2.4.3 Pythagorean Identities

1. The three Pythagorean identities:

$$\begin{aligned}\cos^2 x+\sin^2 &= 1\\ 1+\tan^2 x &= \sec^2 x\\ 1+\cot^2 x &= \csc^2 x\end{aligned}$$

2. Follows directly from the definition of sin and cos using the unit circle!
3. See solution!

2.4.4 Double and Half Angle Formulas

1. $\sin 2x = 2\cos x \sin x$

2. $\cos 2x = \cos^2 x - \sin^2 x = 2\cos^2 x - 1 = 1 - 2\sin^2 x$

3. $\tan 2x = \dfrac{2\tan x}{1 - \tan^2 x}$

4. $\cos^2 x = \frac{1}{2}(1 + \cos 2x)$

5. $\sin^2 x = \frac{1}{2}(1 - \cos 2x)$

6. $\sin \frac{x}{2} = \pm\sqrt{\dfrac{1 - \cos x}{2}}$

7. $\cos \frac{x}{2} = \pm\sqrt{\dfrac{1 + \cos x}{2}}$

8. $\tan \frac{x}{2} = \dfrac{\sin x}{1 + \cos x} = \dfrac{1 - \cos x}{\sin x}$

2.4.5 The Laws of Sines and Cosines

1. $a = 7,\ \beta = \arcsin\left(\frac{3}{14}\sqrt{3}\right) \approx 0.3803$,
   $\gamma = \frac{1}{3}\pi - \beta \approx 0.6669$.

2. $b \approx 13.2443, \alpha \approx 89.8180°, \gamma \approx 28.1820°$

3. $\alpha \approx 33.7133°, \beta \approx 107.0796°, \gamma \approx 39.2072°$

4. $\beta \approx 57.4227°, \gamma = 180° - \alpha - \beta \approx 86.0773°$,
   $c \approx 40.2536$

5. $\beta = \frac{5\pi}{12}, b \approx 5.4641, c = 2\sqrt{6} \approx 4.8990$

2.4.6 Special Values of Trigonometric Functions

1. 0
2. $\frac{1}{2}$
3. -1
4. $\frac{-\sqrt{2}}{2}$
5. $\sqrt{2}$
6. $\frac{\sqrt{3}}{3}$
7. $\frac{1}{512}$

2.4.7 Simplify Trigonometric Expression Involving Inverse Trigonometric Functions

1. $-\pi/4$

2. $\sqrt{3}/2$

3. $\frac{1}{2}$

4. $\frac{1}{\sqrt{1+x^2}}$

5. $\frac{\sqrt{4+x^2}}{2}$

6. $\sqrt{1-x^2}$

7. $\frac{1}{\sqrt{1-x^2}}$

2.4.8 Graphing Trigonometric Functions

1. See Summary!

2. See Summary!

3. See Summary!

4. See Summary!

5. See Summary!

6. See Summary!

7. Note $y = 2\sin\left(\frac{1}{2}x\right)$ oscillates with "half" the frequency of $\sin(x)$ but twice the amplitude!

8. Note that the graph of $y = -\cos\left(x - \frac{\pi}{2}\right)$ is obtained from the graph of $y = \cos(x)$ by reflection on the x-axes and a "phase-shift" of $\frac{\pi}{2}$ (to the right).

9. Note that $y = \sin(x - \pi) - 2$ is obtained from $y = \sin(x)$ by a phase shift of π (to the right) and a vertical shift of 2 downwards.

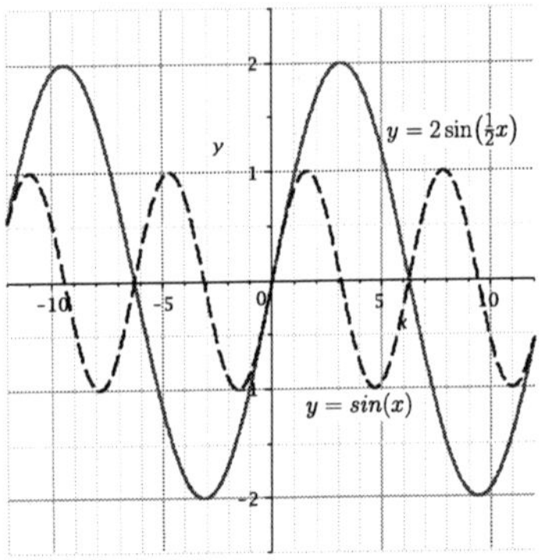

Figure 2.21:

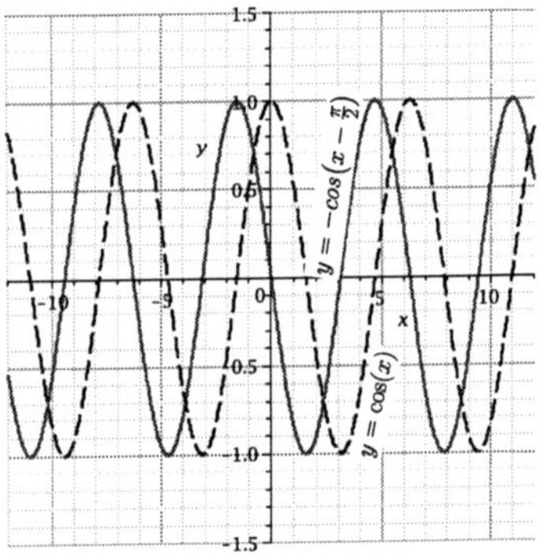

Figure 2.22:

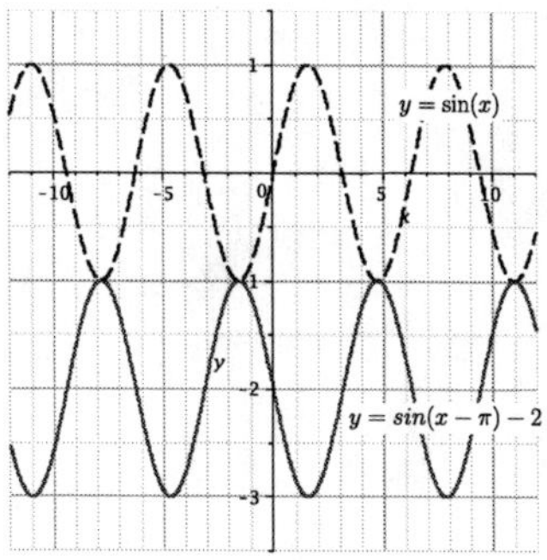

Figure 2.23:

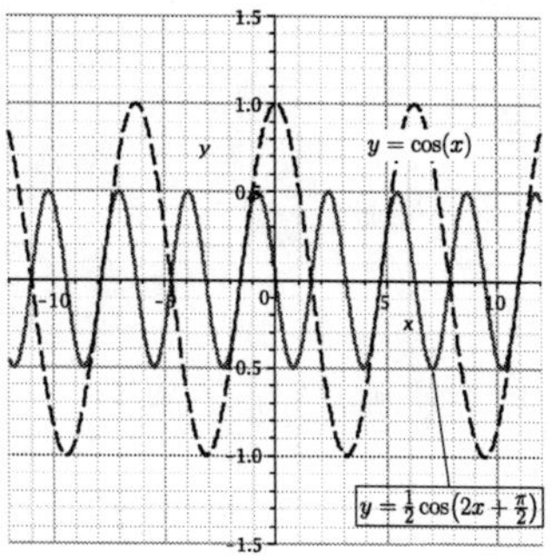

Figure 2.24:

10. Note $y = \frac{1}{2}\cos(2x + \frac{\pi}{2})$ has amplitude $\frac{1}{2}$, phase shift $\frac{\pi}{2}$ (to the left) and frequency 2.

2.4.9 Trigonometric Equations

1. $x = \pi/4(1 + 4n)$ for all $n \in \mathbb{Z}$
2. $x = \pi/6(1 + 12n)$ or $x = \pi/6(5 + 12m)$ for all $n, m \in \mathbb{Z}$
3. $x = \pi/2(4n + 1)$ or $x = \pi/2(4m + 3)$ for all $m, n \in \mathbb{Z}$
4. $x = \pi/2(1 + 4n\pi)$ or $x = -\pi/2(1 - 4m\pi)$ for all $m, n \in \mathbb{Z}$.
5. $x = \frac{5}{3}\pi + 2n\pi$ or $x = \frac{4}{3}\pi + 2m\pi$ for all $m, n \in \mathbb{Z}$
6. $x = \frac{\pi}{4} + 2n\pi$ or $x = -\frac{\pi}{4} + 2m\pi$ for all $m, n \in \mathbb{Z}$
7. Let $\theta := \arccos(-3 + \sqrt{5}) \in [0, \pi]$. Then $x = \theta + 2n\pi$ or $x = 2\pi - \theta + 2m\pi$ for all $m, n \in \mathbb{Z}$.
8. $x = n\pi$ for all $n \in \mathbb{Z}$
9. $x = \arctan(-1 + \sqrt{6}) + n\pi$ or
   $x = \arctan(-1 - \sqrt{6}) + m\pi$ for all $m, n \in \mathbb{Z}$

2.4.10 Converting Cartesian into Polar Coordinates

1. $(1, \frac{\pi}{2})$
2. $(3, \pi)$
3. $(2\sqrt{2}, \frac{7\pi}{4})$
4. $(\sqrt{2x}, \frac{\pi}{4})$
5. $(\sqrt{6}, \frac{5\pi}{3})$
6. $(2, \frac{11\pi}{6})$

2.4.11 Converting Polar into Cartesian Coordinates

1. $(\frac{1}{2}, \frac{\sqrt{3}}{2})$
2. $(-\frac{3\sqrt{2}}{2}, \frac{3\sqrt{2}}{2})$
3. $(-2, 0)$
4. $(\frac{-3}{2}, \frac{3\sqrt{3}}{2})$
5. $(0, -3)$
6. $(-2, 0)$
7. $(\frac{-\sqrt{2}}{2}, \frac{-\sqrt{2}}{2})$
8. $(\frac{-1}{6}, \frac{\sqrt{3}}{6})$

2.4.12 Hyperbolic Functions

1. $\frac{1}{2}\left(e^x - e^{-x}\right)$
2. $\frac{1}{2}\left(e^x + e^{-x}\right)$
3. $\dfrac{e^{2x} - 1}{e^{2x} + 1}$
4. $\dfrac{e^{2x} + 1}{e^{2x} - 1}$
5. $\dfrac{2e^x}{e^{2x} + 1}$
6. $\dfrac{2e^x}{e^{2x} - 1}$

2.4.13 Hyperbolic Identies

1. 1
2. $\sinh(x + y)$
3. $\cosh(x + y)$
4. $\frac{\sinh(2x)}{2}$
5. $\cosh(2x)$
6. $\coth^2(x)$

2.5 Solutions to Selected Exercises

2.5.1 Simplifying Trigonometric Expressions

1. We will use the Double-Angle Identity

$$\sin(2x) = 2\sin x \cos x$$

and the Power Identity

$$\cos^2 x = \tfrac{1}{2}\big(1 + \cos 2x\big)$$

and obtain

$$\begin{aligned} &\sin x \cos x + \cos^2 x = \\ &\quad = \tfrac{1}{2}\sin(2x) + \tfrac{1}{2} + \tfrac{1}{2}\cos(2x) \\ &\quad = \tfrac{1}{2}\big(1 + \cos(2x) + \sin(2x)\big). \end{aligned}$$

2. We will factor out $2\sin x \cos x$ and apply the Double Angle Identities several times:

$$\begin{aligned} &2\sin x \cos^3 x - 2\sin^3 x \cos x = \\ &\quad = 2\sin x \cos x\big(\cos^2 x - \sin^2 x\big) \\ &\quad = \sin(2x)\cos(2x) = \tfrac{1}{2}\sin(4x). \end{aligned}$$

3. We will first factor the numerator using $a^2 - b^2 = (a-b)(a+b)$ and then apply the Pythagorean Identity $1 + \tan^2 x = \sec^2 x$:

$$\begin{aligned} &\frac{\sec^4 x - \tan^4 x}{\sec^2 x + \tan^2 x} = \\ &= \frac{(\sec^2 x - \tan^2 x)(\sec^2 x + \tan^2 x)}{\sec^2 x + \tan^2 x} \\ &= \sec^2 x - \tan^2 x = 1. \end{aligned}$$

4. Simplifying the expression, we obtain

$$\begin{aligned} \frac{2\sin^2 x}{\cos^3 x} \cdot \left(\frac{\cos x}{2\sin x}\right)^2 &= \frac{2\sin^2 x \cos^2}{\cos^3 x\, 4\, \sin^2 x} \\ &= \frac{1}{2\cos x} = \tfrac{1}{2}\sec x \end{aligned}$$

5. We can reduce the fraction by first factoring out $\sin x \cos x$ in the numer-

ator:

$$\begin{aligned}\frac{3\sin x}{\cos^2 x}\cdot\frac{\cos^2 x+\cos x\sin x}{\sin^2 x-\cos^2 x} &=\\ &=\frac{3(\sin x\cos^2 x+\sin^2 x\cos x)}{\cos^2 x(\sin^2 x-\cos^2 x)}\\ &=\frac{3\sin x\cos x(\cos x+\sin x)}{\cos^2 x(\sin x-\cos x)(\sin x+\cos x)}\\ &=\frac{3\sin x}{\cos x(\sin x-\cos x)}\\ &=\frac{3\tan(x)}{\sin x-\cos x}.\end{aligned}$$

6. We first compactify the denominator by applying the Pythagorean Identity $1+\tan^2\theta=\sec^2\theta$. Then we reduce the fraction after rewriting it in terms of sin and cos:

$$\begin{aligned}\frac{2\tan\theta}{1+\tan^2\theta} &= \frac{2\tan\theta}{\sec^2\theta}=\frac{2\sin\theta\cos^2\theta}{\cos\theta}\\ &=2\sin\theta\cos\theta=\sin(2\theta).\end{aligned}$$

7. We separate the fraction and then rewrite $\tan$ in terms of $\sin$ and $\cos$:

$$\begin{aligned}\frac{\tan x-\sin x}{2\tan x} &= \frac{\tan x}{2\tan x}-\frac{\sin x}{2\tan x}\\ &=\frac{1}{2}-\frac{\sin x\cos x}{2\sin x}\\ &=\tfrac{1}{2}-\tfrac{1}{2}\cos x=\tfrac{1}{2}(1-\cos x).\end{aligned}$$

8. With the Addition formula for sin, follows

$$\begin{aligned}&\sin(\alpha+\beta)\sin(\alpha-\beta)=\\ &=\big(\sin\alpha\cos\beta+\sin\beta\cos\alpha\big)\\ &\qquad\cdot\big(\sin\alpha\cos\beta-\sin\beta\cos\alpha\big)\\ &=\sin^2\alpha\cos^2\beta-\sin^2\beta\cos^2\alpha\\ &=\tfrac{1}{2}\big(1-\cos 2\alpha\big)\tfrac{1}{2}\big(1+\cos 2\beta\big)\\ &\qquad-\tfrac{1}{2}\big(1-\cos 2\beta\big)\tfrac{1}{2}\big(1+\cos 2\alpha\big)\end{aligned}$$

from which follows using the abbreviations $A := \cos 2\alpha$ and $B := \cos 2\beta$

$$\begin{aligned}
&= \tfrac{1}{4}\Big((1-A)(1+B) - (1-B)(1+A)\Big)\\
&= \tfrac{1}{4}\Big(1 - A + B - AB - (1 + A - B - AB)\Big)\\
&= \tfrac{1}{4}\Big(1 - A + B - AB - 1 - A + B + AB\Big)\\
&= \tfrac{1}{4}(-2A + 2B) = -\tfrac{1}{2}A + \tfrac{1}{2}B\\
&= -\tfrac{1}{2}\cos 2\alpha + \tfrac{1}{2}\cos 2\beta.
\end{aligned}$$

9. We combine the fractions and reduce after applying the third Pythagorean Identity:

$$\begin{aligned}
\left(\frac{\cot x}{\csc x}\right)^2 + \frac{1}{\csc^2 x} &=\\
&= \frac{\cot^2 x}{\csc^x} + \frac{1}{\csc^2 x}\\
&= \frac{\cot^2 x + 1}{\csc^2 x}\\
&= \frac{\csc^2 x}{\csc^2 x} = 1.
\end{aligned}$$

10. We separate the fraction and obtain

$$\begin{aligned}
\frac{\cos\theta + \sin\theta}{\cos\theta} &= \frac{\cos\theta}{\cos\theta} + \frac{\sin\theta}{\cos\theta}\\
&= 1 + \tan\theta.
\end{aligned}$$

11. Applying the Pythagorean identities $1+\tan^2 x = \sec^2 x$ and $1 = \cos^2 x + \sin^2 x$, we obtain

$$\begin{aligned}
(\cos^2 x) \cdot \frac{1+\tan^2 x}{1-\cos^2 x} &=\\
&= \frac{\cos^2 x \sec^2 x}{\sin^2 x}\\
&= \frac{\cos^2 x}{\sin^2 x \cos^2 x}\\
&= \frac{1}{\sin^2 x} = \csc^2 s.
\end{aligned}$$

2.5.2 Sum and Difference Formulas for Sine and Cosine

1. $\sin(x+y) = \sin x \cos y + \sin y \cos x$

2. $\cos(x+y) = \cos x \cos y - \sin x \sin y$

3. $\sin(x - y) = \sin x \cos y - \sin y \cos x$

4. $\cos(x - y) = \cos x \cos y + \sin x \sin y$

2.5.3 Pythagorean Identities

1. The three Pythagorean identities:

$$\begin{aligned} \cos^2 x + \sin^2 &= 1 \\ 1 + \tan^2 x &= \sec^2 x \\ 1 + \cot^2 x &= \csc^2 x \end{aligned}$$

2. Follows directly from the definition of sin and cos using the unit circle!

3. Using the First Pythagorean identity, we derive

$$\begin{aligned} 1 + \tan^2(x) &= 1 + \frac{\sin^2(x)}{\cos^2(x)} \\ &= \frac{\cos^2(x)}{\cos^2(x)} + \frac{\sin^2(x)}{\cos^2(x)} \\ &= \frac{\cos^2(x) + \sin^2(x)}{\cos^2(x)} \\ &= \frac{1}{\cos^2(x)} = 1 \end{aligned}$$

2.5.4 Double and Half Angle Formulas

1. $\sin 2x = 2 \cos x \sin x$

2. $\cos 2x = \cos^2 x - \sin^2 x = 2\cos^2 x - 1 = 1 - 2\sin^2 x$

3. $\tan 2x = \dfrac{2 \tan x}{1 - \tan^2 x}$

4. $\cos^2 x = \frac{1}{2}\big(1 + \cos 2x\big)$

5. $\sin^2 x = \frac{1}{2}\big(1 - \cos 2x\big)$

6. $\sin \frac{x}{2} = \pm\sqrt{\dfrac{1 - \cos x}{2}}$

7. $\cos \frac{x}{2} = \pm\sqrt{\dfrac{1 + \cos x}{2}}$

8. $\tan \frac{x}{2} = \dfrac{\sin x}{1 + \cos x} = \dfrac{1 - \cos x}{\sin x}$

2.5.5 The Laws of Sines and Cosines

1. We need to compute a, β and γ. Clearly, $\gamma = \pi - \alpha - \beta = \pi - \frac{2}{3}\pi - \beta = \frac{1}{3}\pi = \beta$. a can be computed using the law of consine:

$$\begin{aligned} a^2 &= b^2 + c^2 - 2bc\cos\alpha \\ &= 9 + 25 - 2\cdot 3\cdot 5\cdot \cos\left(\tfrac{2}{3}\pi\right) \\ &= 34 - 30\left(\cos^2\tfrac{\pi}{3} - \sin^2\tfrac{\pi}{3}\right) \\ &= 34 - 30\left(\tfrac{1}{4} - \tfrac{3}{4}\right) = 49 \end{aligned}$$

hence

$$a = \sqrt{49} = 7.$$

To compute β, we will use the law of sine:

$$\frac{\sin\beta}{\sin\alpha} = \frac{3}{7},$$

hence

$$\sin\beta = \tfrac{3}{7}\sin\alpha = \tfrac{3}{7}\cdot\tfrac{1}{2}\sqrt{3} = \tfrac{3}{14}\sqrt{3}$$

and therefore

$$\beta = \arcsin\left(\tfrac{3}{14}\sqrt{3}\right).$$

2. We need to compute b, α and γ. Clearly, $\gamma = 180° - 62° - \alpha = 118° - \alpha$. b can be computed using the law of cosine:

$$\begin{aligned} b^2 &= a^2 + c^2 - 2ac\cos\beta \\ &= 225 + 49 - 2\cdot 15\cdot 7\cdot\cos 62° \\ &= 274 - 210\cos 62° \end{aligned}$$

hence

$$b = \sqrt{274 - 210\cos 62°} \approx 13.2443.$$

Since, by the law of sine,

$$\frac{\sin\alpha}{\sin\beta} = \frac{a}{b}$$

it follows that

$$\sin\alpha = \frac{15}{b}\sin 62°,$$

hence

$$\alpha = \arcsin\left(\frac{15}{b}\sin 62^\circ\right) \approx 89.8180$$

and thus

$$\gamma = (180 - 62 - \alpha)^\circ \approx 28.1820^\circ.$$

3. We use the Law of Cosines to first compute γ. Since $c^2 = a^2 + b^2 - 2ab\cos\gamma$, we obtain

$$\gamma = \arccos\left(\frac{a^2 + b^2 - c^2}{2ab}\right) \approx 39.2072^\circ.$$

Similarly, we obtain

$$\alpha = \arccos\left(\frac{b^2 + c^2 - a^2}{2bc}\right) \approx 33.7133^\circ.$$

Hence, $\beta = 180^\circ - \alpha - \gamma \approx 107.0796^\circ$.

4. β can be easily computed using the Law of Sines, indeed

$$\sin\beta = \frac{b}{a}\sin\alpha$$

which implies

$$\beta = \arcsin\left(\frac{b}{a}\sin\alpha\right) \approx 57.4227$$

and therefore, $\gamma = 180^\circ - \alpha - \beta \approx 86.0773$. With the Law Cosines, we can now compute c and obtain

$$c = \sqrt{b^2 + a^2 - 2ba\cos(\gamma)} \approx 40.2536.$$

5. Clearly, $\beta = \pi - \alpha - \gamma = \pi(-\frac{1}{4}\pi - \frac{1}{3}\pi) = \frac{5}{12}\pi$. c and b can becomputed using the Law of Sines:

$$\frac{c}{a} = \frac{\sin\gamma}{\sin\alpha}$$

and therefore

$$c = \frac{a\sin\gamma}{\sin\alpha} = 2\sqrt{6} \approx 4.8990$$

similarly

$$\frac{b}{a} = \frac{\sin\beta}{\sin\alpha}$$

and thus

$$b = \frac{a\sin\beta}{\sin\alpha} \approx 5.4641.$$

2.5.6 Special Values of Trigonometric Functions

2.5.7 Simplify Trigonometric Expression Involving Inverse Trigonometric Functions

2.5.8 Graphing Trigonometric Functions

1. See Summary!
2. See Summary!
3. See Summary!
4. See Summary!
5. See Summary!
6. See Summary!
7. See corresponding answer!
8. See corresponding answer!
9. See the corresponding answer!
10. See the corresponding answer!

2.5.9 Trigonometric Equations

1. Consulting the unit circle definition of sin, we find that $\sin(2x) = 1 \Leftrightarrow 2x = \frac{\pi}{2} + 2n\pi$ for all $n \in \mathbb{Z}$. Hence, $x = \frac{\pi}{4}(1 + 4\pi)$ for all $n \in \mathbb{Z}$.

2. Consulting the figure below, we find that $\sin(x) = \frac{1}{2} \Leftrightarrow x = \frac{\pi}{6} + 2n\pi$ or x= $\pi - \frac{\pi}{6} + 2m\pi$ for all $m, n \in \mathbb{Z}$. Simplifying these expressions, we obtain that $x = \frac{\pi}{6}(1 + 12n)$ or $x = \frac{\pi}{6}(5 + 12m)$ for all $m, n \in \mathbb{Z}$.

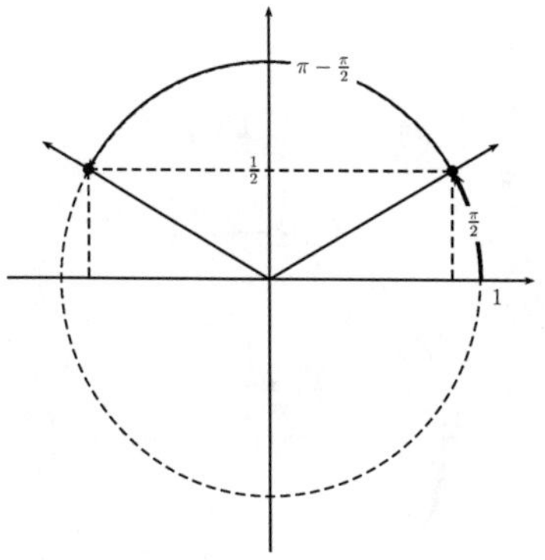

Figure 2.25:

3. Consulting the figure below, we find that $\tan(x/2) = 1 \Leftrightarrow \frac{1}{2}x = \frac{\pi}{4} + n\pi$ or $\frac{1}{2}x = \pi - \frac{\pi}{4} + m\pi$ for all $m, n \in \mathbb{Z}$. Simplifying these expression, we obtain $x = \pi/2(4n + 1)$ or $x = \pi/2(4m + 3)$.

4. Squaring the equation, we obtain $|\sin(x)| = 1$ which is equivalent to $\sin(x) = \pm 1$. Consulting the definition of sin using the unit circle, we find that $x = \frac{\pi}{2} + 2n\pi$ or $x = -\frac{\pi}{2} + 2m\pi$, which simplifies to $x = \pi/2(1 + 4n\pi)$ or $x = -\pi/2(1 - 4m\pi)$ for all $m, n \in \mathbb{Z}$.

5. Consulting the figure below, we find that $\sin 2x = -\frac{1}{2}\sqrt{3}$ is equivalent to $x = \frac{3}{2}\pi - \frac{\pi}{6} + 2n\pi$ or $x = \frac{3}{2}\pi + \frac{\pi}{6} + 2m\pi$ for all $m, n \in \mathbb{Z}$, which simplifies to $x = \frac{4}{3}\pi + 2n\pi$ or $x = \frac{5}{3}\pi + 2m\pi$ for $m, n \in \mathbb{Z}$.

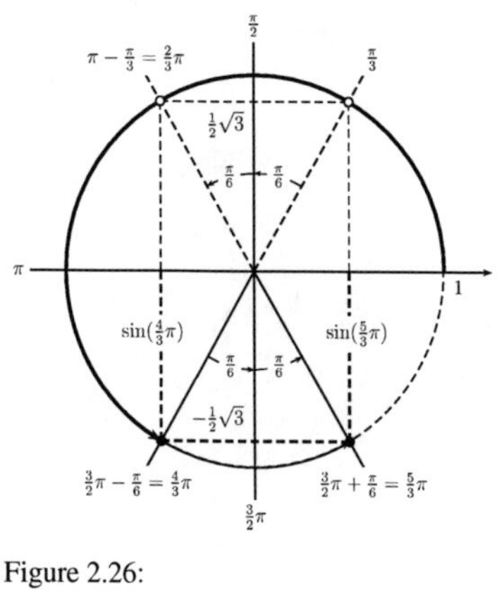

Figure 2.26:

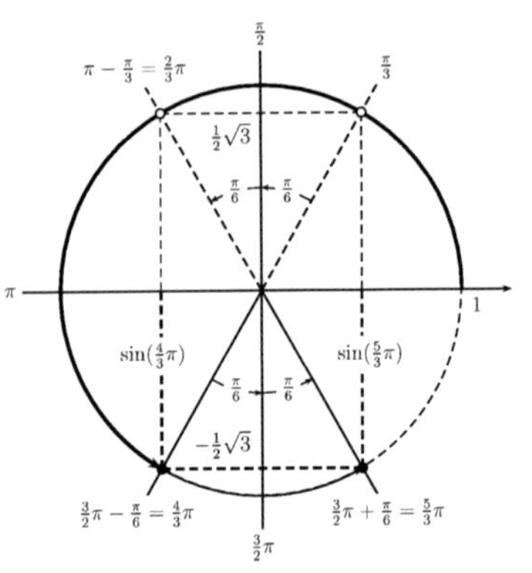

Figure 2.27:

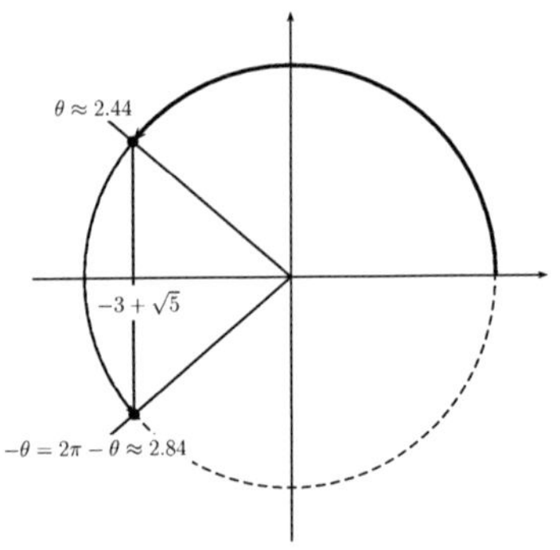

Figure 2.28:

7. The substitution $y = \cos x$ reduces the given equation to the quadratic equation $y^2 + 6y + 4 = 0$ which using the "quadratic formula" yields the solutions

$$\begin{aligned} y &= \frac{-6 \pm \sqrt{36 - 16}}{2} = \frac{-6 \pm 2\sqrt{5}}{2} \\ &= -3 \pm \sqrt{5}. \end{aligned}$$

We now solve the remaining equations $\cos x = y$. Clearly, only $y = -3+\sqrt{5}$ leads to an equation that can be solved in $\mathbb{R}$. Let $\theta := \arccos(-3+\sqrt{5})$. Note that $\theta \in [0, \pi]$. Consulting the figure below (or remembering that cos is an even function!), we see that also $-\theta = 2\pi - \theta$ solves $\cos x = -3 + \sqrt{5}$. Hence, $x = \theta + 2n\pi$ or $x = 2\pi - \theta + 2m\pi$ for all $m, n \in \mathbb{Z}$.

8. Using the Double-Angle Identity $\sin 2x = \sin x \cos x$, we see that $\sin 2x \cos x + \sin x = 0$ is equivalent to

$$\begin{aligned} 2 \sin x \cos^2 x + \sin x &= 0 \\ \sin x (2\cos^2 x + 1) &= 0 \end{aligned}$$

which splits into the equations $\sin x = 0$ and $\cos^2 x = -\frac{1}{2}$. Clearly, the last equation does not have a solution in $\mathbb{R}$. Hence, $x = n\pi$ for all $n \in \mathbb{Z}$.

9. The identity $\sec^2 x = 1 + \tan^2 x$ reduces the given equation to $\tan^2 x + 2\tan x - 5 = 0$ which reduces further to the quadratic equation $y^2 + 2y - 5 = 0$ by applying the substitution $y = \tan x$. With the "quadratic formula", we obtain

$$y = \frac{-2 \pm \sqrt{4 + 20}}{2} = -1 \pm \sqrt{6}$$

The remaining equations $\tan x = -1 \pm \sqrt{6}$ clearly have the solutions

$$\begin{aligned} x &= \arctan(-1 + \sqrt{6}) + n\pi \\ x &= \arctan(-1 - \sqrt{6}) + m\pi \end{aligned}$$

for all $m, n \in \mathbb{Z}$.

2.5.10 Converting Cartesian into Polar Coordinates

2.5.11 Converting Polar into Cartesian Coordinates

2.5.12 Hyperbolic Functions

2.5.13 Hyperbolic Identies

3

DIFFERENTIAL CALCULUS

3.1 Definitions and Results

3.1.1 Limits of Functions

The notion of limit in its different forms is the central notion of calculus. We will first consider the notion of the limit of a function.

Definition 3.1 (Two-sided Limit)
Suppose the function f is defined in a neighborhood of $c \in \mathbb{R}$. We say that *f has limit L as x approaches c* and write

$$\lim_{x \to c} f(x) = L$$

if and only if for all $\varepsilon > 0$ there exists a $\delta > 0$ such that for all $x \in D_f$

$$0 < |x - c| < \delta \Rightarrow |f(x) - L| < \varepsilon$$

(D_f denotes the domain of f).

Remark 3.1 (Uniqueness of Limit)
If f has a limit at c, this limit is unique. The same is true for half-sided limits.

Definition 3.2 (Left-sided Limit)
Suppose the function f is defined in a half-neighborhood of the form (a, c) of $c \in \mathbb{R}$. We say that f *has limit L as x approaches c from the left* and write

$$\lim_{x \to c^-} f(x) = L$$

if and only if for all $\varepsilon > 0$ there exists a $\delta > 0$ such that for all $x \in D_f$

$$0 < c - x < \delta \Rightarrow |f(x) - L| < \varepsilon$$

Definition 3.3 (Right-sided Limit)
Suppose the function f is defined in a half-neighborhood of the form (c, b) of $c \in \mathbb{R}$. We say that f *has limit L as x approaches c from the right* and write

$$\lim_{x \to c^+} f(x) = L$$

if and only if for all $\varepsilon > 0$ there exists a $\delta > 0$ such that for all $x \in D_f$

$$0 < x - c < \delta \Rightarrow |f(x) - L| < \varepsilon$$

Definition 3.4 (Limits at Infinity)

1. Suppose the function f is defined on an interval of the form (a, ∞). We say that f *has limit L as x approaches ∞* and write

   $$\lim_{x \to \infty} f(x) = L$$

   if and only if for all $\varepsilon > 0$ there exists a $B > 0$ such that for all $x \in (a, \infty)$

   $$x > B \Rightarrow |f(x) - L| < \varepsilon.$$

   The line $y = L$ is called a *horizontal asymptote* of f.

2. Suppose the function f is defined on an interval of the form $(-\infty, b)$. We say that f *has limit L as x approaches $-\infty$* and write

   $$\lim_{x \to -\infty} f(x) = L$$

   if and only if for all $\varepsilon > 0$ there exists a $B > 0$ such that for all $x \in (-\infty, b)$

   $$x < -B \Rightarrow |f(x) - L| < \varepsilon.$$

   The line $y = L$ is called a *horizontal asymptote* of f.

Definition 3.5 (Infinite Limits)

1. Suppose the function f is defined in a neighborhood of $c \in \mathbb{R}$. We say that f *approaches* ∞ *as* x *approaches* c and write
$$\lim_{x \to c} f(x) = \infty$$
if and only if for all $B > 0$ there exists a $\delta > 0$ such that for all $x \in D_f$
$$0 < |x - c| < \delta \Rightarrow f(x) > B.$$
The line $x = c$ is called a *vertical asymptote.*

2. Suppose the function f is defined in a neighborhood of $c \in \mathbb{R}$. We say that f *approaches* $-\infty$ *as* x *approaches* c and write
$$\lim_{x \to c} f(x) = -\infty$$
if and only if for all $B > 0$ there exists a $\delta > 0$ such that for all $x \in D_f$
$$0 < |x - c| < \delta \Rightarrow f(x) < -B.$$
The line $x = c$ is called a *vertical asymptote.*

Theorem 3.1 (Half- and Two-sided Limits)

1. *If* $\lim_{x \to c^-} f(x) = \lim_{x \to c^+} f(x) = L$, *then* $\lim_{x \to c} f(x) = L$.
2. *If* $\lim_{x \to c} f(x) = L$, *then* $\lim_{x \to c^-} f(x) = \lim_{x \to c^+} f(x) = L$.
3. *If* $\lim_{x \to c^-} f(x)$ *or* $\lim_{x \to c^+} f(x)$ *fail to exist, then* $\lim_{x \to c} f(x)$ *fails to exist.*
4. *If* $\lim_{x \to c^-} f(x) \neq \lim_{x \to c^+} f(x)$, *then* $\lim_{x \to c} f(x)$ *fails to exist.*

Theorem 3.2 (Limits of Elementary Functions I)

1. $\lim_{x \to c} k = k$.
2. $\lim_{x \to c} x = c$
3. *If* p *is a polynomial, i.e. a function of the form* $p(x) = a_0 + a_1x + a_2x^2 + \cdots + a_nx^n$, *then* $\lim_{x \to c} p(x) = p(c)$.
4. *If* r *is a rational function (quotient of polynomials), and* $c \in D_r$ *(i.e. not a zero of the denominator of* r*) then* $\lim_{x \to c} r(x) = r(c)$.

Theorem 3.3 (Limits of Elementary Functions II)

1. $\lim_{x \to c} \sqrt[n]{x} = \sqrt[n]{c}$, ($x \geq 0$ if n is even).
2. $\lim_{x \to c} \sin x = \sin c$.
3. $\lim_{x \to c} \cos x = \cos c$
4. $\lim_{x \to c} e^x = e^c$.
5. $\lim_{x \to c} \ln x = \ln c, \quad x > 0$.
6. $\lim_{x \to 0} \dfrac{\sin x}{x} = 1$.
7. $\lim_{x \to 0} \dfrac{\cos x - 1}{x} = 0$
8. $\lim_{x \to \pm\infty} \dfrac{1}{x} = 0$
9. $\lim_{x \to 0^+} \dfrac{1}{x} = \infty, \quad \lim_{x \to 0^-} \dfrac{1}{x} = -\infty$.
10. $\lim_{x \to \infty} e^x = \infty, \quad \lim_{x \to -\infty} e^x = 0$.
11. $\lim_{x \to \infty} \ln x = \infty, \quad \lim_{x \to 0^+} \ln x = -\infty$.
12. $\lim_{x \to \infty} \tan^{-1} x = \dfrac{\pi}{2}$,
    $\lim_{x \to -\infty} \tan^{-1} x = -\dfrac{\pi}{2}$.

3.1.2 Techniques for Computing Limits

Theorem 3.4 (Limit Theorem)
Suppose the functions f and g have limits at c and suppose $k \in \mathbb{R}$, then

1. $\lim\limits_{x\to c}\big(f(x)+g(x)\big) = \lim\limits_{x\to c} f(x) + \lim\limits_{x\to c} g(x)$.
2. $\lim\limits_{x\to c}\big(kf(x)\big) = k\lim\limits_{x\to c} f(x)$.
3. $\lim\limits_{x\to c}\big(f(x)g(x)\big) = \lim\limits_{x\to c} f(x)\lim\limits_{x\to c} g(x)$.
4. $\lim\limits_{x\to c}\dfrac{f(x)}{g(x)} = \dfrac{\lim\limits_{x\to c} f(x)}{\lim\limits_{x\to c} g(x)}$, *provided that* $\lim\limits_{x\to c} g(x) \neq 0$.
5. *If* $\lim\limits_{x\to c} f(x) \neq 0$ *and* $\lim\limits_{x\to c} g(x) = 0$, *then* $\lim\limits_{x\to c}\dfrac{f(x)}{g(x)}$ *does not exist.*

Example 3.1 (Limits of Polynomials)
Let

$$p(x) = a_0 + a_1x + a_2x^2 + \cdots + a_nx^n = \sum_{k=0}^{n} a_kx^k$$

be a polynomial and $c \in \mathbb{R}$. Find $\lim\limits_{x\to c} p(x)$.

SOLUTION: By first applying the Summation and Scalar-Multiple Rule for Limits (Theorem 3.4:1 and 2), the limit reduces to

$$\lim_{x\to c} p(x) = \lim_{x\to c}\left(\sum_{k=0}^{n} a_kx^k\right) = \sum_{n=0}^{n} a_k \lim_{x\to c} x^k$$

which with the Product Rule (Theorem 3.4:3) and Theorem 3.2:2 yields

$$= \sum_{k=0}^{n} a_k\left(\lim_{x\to c} x\right)^k = \sum_{k=0}^{n} a_k\,(c)^k = p(c)$$

Thus, the limit of the polynomial p at c is the function value $p(c)$. In other words, polynomials are continuous on $\mathbb{R}$. ◁

Example 3.2 (Limits of Rational Functions)
Let

$$r(x) = \frac{p(x)}{q(x)}$$

be a rational function, where p and q are polynomials. Let c be an element of the domain of r, i.e. $c \in D_r := \{x \in \mathbb{R} \mid q(x) \neq 0\}$. Find $\lim\limits_{x\to c} r(x)$.

SOLUTION: Since by the previous example, polynomials are continuous, $\lim_{x\to c} p(x) = p(c)$ and $\lim_{x\to c} q(x) = q(c) \neq 0$. Hence, by the Quotient Rule for Limits (Theorem 3.4:4), we have

$$\lim_{x\to c} \frac{p(x)}{q(x)} = \frac{\lim_{x\to c} p(x)}{\lim_{x\to c} q(x)} = \frac{p(c)}{q(c)} = r(c).$$

Thus, the limit of the rational function r at an element c of its domain is the function value $r(c)$. In other words, rational functions are continuous on their domain. ◁

Theorem 3.5 (Substitution Rule)
If $\lim_{x\to x_0} f(x) = y_0$ *and* $\lim_{y\to y_0} g(y) = g(y_0)$, *then*

$$\lim_{x\to x_0} g(f(x)) = g\Big(\lim_{x\to x_0} f(x)\Big) = g(y_0)$$

Example 3.3
Evaluate the limit

$$\lim_{x\to 1} \frac{3x-6}{\sqrt[3]{x-2}}$$

SOLUTION: We first reduce the argument of the limit by cancellation

$$\lim_{x\to 1} \frac{3x-6}{\sqrt[3]{x-2}} = \lim_{x\to 1} \frac{3(x-2)}{(x-2)^{2/3}} = \lim_{x\to 1} 3(x-2)^{2/3}$$

then apply the Scalar-Multiple and the Substitution Rule for limits (Theorem 3.4:2 and Theorem 3.5) which yields

$$= 3\lim_{x\to 1}(x-2)^{2/3} = 3\left(\lim_{x\to 1}(x-2)\right)^{2/3}$$

and, since the argument of the remaining limit is a polynomial, its value equals $1-2=-1$. Hence, we obtain

$$= 3(1-2)^{2/3} = 3$$

◁

Corollary 3.1
Suppose the function f has a limit at c and $n \in \mathbb{N}$, then

1. $\lim_{x\to c} [f(x)]^n = \left[\lim_{x\to c} f(x)\right]^n$
2. $\lim_{x\to c} \sqrt[n]{f(x)} = \sqrt[n]{\lim_{x\to c} f(x)}$,
   provided $f(x) \geq 0$ in a neigborhood of c, if n is even.

Theorem 3.6 (Inequalities)
If f and g have limits at x_0 and if $f(x) \leq g(x)$ for all x in an open interval containing x_0 then

$$\lim_{x \to x_0} f(x) \leq \lim_{x \to x_0} g(x).$$

Theorem 3.7 (Squeezing Theorem)
If $\lim_{x \to c} f(x) = \lim_{x \to c} g(x) = L$ and if $f(x) \leq h(x) \leq g(x)$ for all x in an open interval containing c, then $\lim_{x \to c} h(x) = L$.

Example 3.4
Suppose $\lim_{x \to c} f(x) = 0$ and suppose that $|g(x)| \leq B$ for all x in an open interval containing c. Show that

$$\lim_{x \to c} f(x)g(x) = 0$$

SOLUTION: By assumption, $|g(x)| \leq B$ for all x in some open interval I containing c. Since $|f(x)| \geq 0$ for all x, we can multiply the inequality by $|f(x)|$ and obtain by Theorem 1.2:4(e)ii

$$|f(x)g(x)| = |f(x)||g(x)| \leq |f(x)|A \quad \text{for all } x \in I,$$

which, by Lemma 1.6:1, is equivalent to

$$-|f(x)|A \leq f(x)g(x) \leq |f(x)|A \quad \text{for all } x \in I.$$

Applying the Scalar Multiple Rule for limits (Theorem 3.4:2) and the assumption $\lim_{x \to c} f(x) = 0$, which is equivalent to $\lim_{x \to c} |f(x)| = 0$ to the outside terms of the double-inequality, we obtain

$$\lim_{x \to c} \big(-|f(x)|A\big) = (-A) \lim_{x \to c} |f(x)| = 0 = A \lim_{x \to c} |f(x)| = \lim_{x \to c} \big(A|f(x)|\big)$$

Therefore, we can conclude with the Squeeze Play (Theorem 3.7) that

$$\lim_{x \to c} f(x)g(x) = 0$$

◁

Figure 3.1: *Guillaume de l'Hôpital (1661–1704)*

Theorem 3.8 (L'Hôpital's Rules)

1. "$\frac{0}{0}$": *If f and g are differentiable in a neighborhood of the point $c \in \mathbb{R} \cup \{\pm\infty\}$, $\lim_{\to c} f(x) = \lim_{\to c} g(x) = 0$ and if*

$$\frac{\lim\limits_{x \to c} f'(x)}{\lim\limits_{x \to c} g'(x)}$$

exists then

$$\frac{\lim\limits_{x \to c} f(x)}{\lim\limits_{x \to c} g(x)} = \frac{\lim\limits_{x \to c} f'(x)}{\lim\limits_{x \to c} g'(x)}$$

2. "$\frac{\infty}{\infty}$": *If f and g are differentiable in a neighborhood of the point $c \in \mathbb{R} \cup \{\pm\infty\}$, $\lim_{\to c} f(x) = \pm \lim_{\to c} g(x) = \pm\infty$ and if*

$$\frac{\lim\limits_{x \to c} f'(x)}{\lim\limits_{x \to c} g'(x)}$$

exists then

$$\frac{\lim\limits_{x \to c} f(x)}{\lim\limits_{x \to c} g(x)} = \frac{\lim\limits_{x \to c} f'(x)}{\lim\limits_{x \to c} g'(x)}$$

Example 3.5
Evaluate the limit $\lim\limits_{x \to 0} \dfrac{\tan \pi x}{e^x - 1}$

SOLUTION: Since $\lim_{x \to 0} \tan \pi x = \lim_{x \to 0}(e^x - 1) = 0$ and the numerator and denominator functions are differentiable, we can apply L'Hôpital's Rule (Theorem 3.8) to the given limit and differentiate the numberator and the denominator functions separately. Clearly, the resulting limit can now be evaluated with the Limit Theorem (Theorem 3.4) in conjuction with Theorem 3.3:

$$\lim_{x \to 0} \frac{\tan \pi x}{e^x - 1} = \lim_{x \to 0} \frac{\pi \sec^2 \pi x}{e^x} = \frac{\pi \left(\lim\limits_{x \to 0} \sec \pi x\right)^2}{\lim\limits_{x \to 0} e^x} = \frac{\pi \cdot 1^2}{1} = \pi$$

◁

3.1.3 Extrema

Definition 3.6 (Absolute or Global Extrema)
Let the function f be defined on the set D_f and let $c \in D_f$. We say that the function f has a

1. *(global or absolute) maximum* at c if $f(x) \leq f(c)$ for all $x \in D_f$; $f(c)$ is then called the *maximum value* of f.
2. *(global or absolute) minimum* at c if $f(x) \geq f(c)$ for all $x \in D_f$; $f(c)$ is then called the *minimum value* of f.
3. *(global) extremum* at c if f has a (global) maximum or minimum at c.

Definition 3.7 (Relative or Local Extrema)
Let the function f be defined on the set D_f and let $c \in D_f$. We say that the function f has a

1. *local or relative maximum* at c if there is an interval (a, b) containing c such that $f(x) \leq f(c)$ for all $x \in (a, b)$
2. *local or relative minimum* at c if there is an interval (a, b) containing c such that $f(x) \geq f(c)$ for all $x \in (a, b)$.
3. *local or relative extremum* at c if f has local or relative maximum or minimum at c.

3.1.4 Continuity

Definition 3.8 (Continuity)
A function f is *continuous at* c if and only if for any $\varepsilon > 0$ there is a $\delta > 0$ such that for all $x \in D_f$ (D_f denotes the domain of f)

$$|x - c| < \delta \Rightarrow |f(x) - f(c)| < \varepsilon$$

If $S \subseteq D_f$ is a subset of the domain D_f of f, then we say that f is *continuous on* S if and only if f is continuous at every point $c \in S$.

Theorem 3.9 (Limit Description of Continuity)
If f is an interior point of D_f then

$$f \text{ is continuous at } c \Leftrightarrow \lim_{x \to c} f(x) = f(c)$$

Theorem 3.10 (Continuity and Algebraic Operations)
If f and g are continuous at c, then

1. *$f + g$ is continuous at c*
2. *$f \cdot g$ is continuous at c*
3. *$\alpha \cdot f$ is continuous at c for any $\alpha \in \mathbb{R}$*
4. *f/g is continuous at c provided $g(c) \neq 0$*
5. *If f is continuous at c and g is continuous at $f(c)$ then the composition $g \circ f$ is continuous at c.*

Theorem 3.11 (Continuity of Elementary Functions)
The following classes of functions are continuous on their domain:

1. *Polynomials,*
2. *Rational Functions,*
3. *Root Functions,*
4. *Trigonometric Functions,*
5. *Inverse Trigonometric Functions,*
6. *Exponential Functions,*
7. *Inverse Trigonometric Functions.*

Definition 3.9 (Discontinuities)
Suppose the function f is defined on the set D and c is an "adherent point" of D. We say that f is *discontinuous* at c if f is not continuous at c. More specifically: The point c is called

1. a *removable discontinuity* of f, if $\lim_{x \to c} f(x)$ exists (and is different from $f(c)$ or $f(c)$ is not defined).
2. a *jump discontinuity* of f, if $\lim_{x \to c^-} f(x)$ and $\lim_{x \to c^+} f(x)$ exist (and are distinct),
3. an *infinite discontinuity* of f, if (at least) one of the half-sided limits is infinite.

Theorem 3.12 (Extreme Value Theorem)
If the function f is continuous on the closed interval $[a, b]$, then there exist numbers c_1 and c_2 in $[a, b]$ such that f has a maximum at c_1 and a minimum at c_2.

Theorem 3.13 (Intermediate Value Theorem)
If the function f is continuous on the closed interval $[a, b]$ and d is any value between the minimum and maximum value of f then there exists a number $c \in [a, b]$ such that $f(c) = d$.

Corollary 3.2 (Existence of Zeros)
If f is continuous on $[a, b]$ and $f(a) \cdot f(b) < 0$, i.e. $f(a)$ and $f(b)$ have different sign, then there exists $c \in (a, b)$ such that $f(c) = 0$, i.e. f has a zero between a and b.

3.1.5 Differentiability

Definition 3.10 (Differentiable, Derivative)
Suppose the function f is defined in an open interval containing the point c. We say that f is *differentiable* at c if the limit

$$\lim_{x \to c} \frac{f(x) - f(c)}{x - c}$$

exists. If the limit exists, it is called the *derivative* of f at c and denoted by $f'(c)$.
If f is differentiable at all $c \in S \subseteq D_f$, then we say, f is *differentiable on* the set S. If f is differentiable on its domain, we simply say f is *differentiable.*

Theorem 3.14 (Differentiablility and Continuity)
If the function f is differentiable at c then it is continuous at c.

Definition 3.11
Given a differentiable function $y = f(x)$, the *differential of* x is denoted by dx, the *differential of* y is defined by

$$df = dy := f'(x)dx.$$

Theorem 3.15 (Differentiation Rules)

1. Summation Rule: *Suppose f and g are differentiable at x then $f+g$ is differentiable at x and*

$$(f+g)'(x) = f'(x) + g'(x)$$

2. Scalar Multiple Rule: *Suppose f is differentiable at x and $\alpha \in \mathbb{R}$, then αf is differentiable at x and*

$$(\alpha f)'(x) = \alpha f'(x)$$

3. Product Rule: *Suppose f and g are differentiable at x then $f \cdot g$ is differentiable at x and*

$$(fg)'(x) = f'(x)g(x) + f(x)g'(x)$$

4. Quotient Rule: *Suppose f and g are differentiable at x and $g(x) \neq 0$, then f/g is differentiable at x and*

$$\left(\frac{f}{g}\right)'(x) = \frac{f'(x)g(x) - f(x)g'(x)}{\big(g(x)\big)^2}$$

5. The Chain Rule: *If the function f is differentiable at x and the function g is differentiable at $f(x)$, then the compositon $g \circ f$ is differentiable at x and*

$$(g \circ f)'(x) = g'\big(f(x)\big) \cdot f'(x)$$

6. The Inverse Function Rule: *If f is differentiable with inverse function f^{-1} then f^{-1} is differentiable at x if $f'\big(f^{-1}(x)\big) \neq 0$ and*

$$\left(f^{-1}\right)'(x) = \frac{1}{f'\big(f^{-1}(x)\big)}$$

Theorem 3.16 (Derivatives of Elementary Functions)

1. $(x^r)' = rx^{r-1}$, $(r \in \mathbb{R})$
2. $\sin' x = \cos x$,
   $\cos' x = -\sin x$
3. $\tan' x = \sec^2 x$,
   $\cot' x = -\csc^2 x$
4. $\sec' x = \sec x \tan x$,
   $\csc' x = -\csc x \cot x$
5. $(e^x)' = e^x$,
   $(a^x)' = a^x \ln a$
6. $\ln' x = \frac{1}{x}$,
   $\log'_a x = \frac{1}{x \ln a}$
7. $\sinh' x = \cosh x$,
   $\cosh' = \sinh x$
8. $\tanh' x = \operatorname{sech}^2 x$,
   $\coth' x = -\operatorname{csch}^2 x$
9. $\operatorname{sech}' x = -\operatorname{sech} x \tanh x$,
   $\operatorname{csch}' x = -\operatorname{csch} x \coth x$
10. $\arcsin' x = \frac{1}{\sqrt{1-x^2}}$,
    $\arccos' x = \frac{-1}{\sqrt{1-x^2}}$
11. $\arctan' x = \frac{1}{1+x^2}$,
    $\operatorname{arccot}' x = \frac{-1}{1+x^2}$
12. $\operatorname{arcsec}' x = \frac{1}{|x|\sqrt{x^2-1}}$,
    $\operatorname{arccsc}' x = \frac{-1}{|x|\sqrt{x^2-1}}$

Theorem 3.17 (Rolle's Theorem)
If the function f is continuous on the interval $[a, b]$, differentiable in the interval (a, b) and $f(a) = f(b)$, then there exists at least one point $c \in (a, b)$ such that $f'(c) = 0$.

Figure 3.2: *Michel Rolle (1655–?)*

Theorem 3.18 (Mean Value Theorem)
If the function f is continuous on the interval $[a, b]$ and differentiable on the interval (a, b), then there exist at least one point $c \in (a, b)$ such that

$$f'(c) = \frac{f(b) - f(a)}{b - a}$$

or equivalently

$$f(b) - f(a) = f'(c)(b - a).$$

3.1.6 Applications of Differential Calculus

Theorem 3.19 (Characterization of Constancy)

1. *Suppose f is differentiable on (a, b). If $f'(x) = 0$ for all $x \in (a, b)$ then f is constant on (a, b).*
2. *Suppose f and g are differentiable on (a, b). If $f'(x) = g'(x)$ for all $x \in (a, b)$ then $f(x) = g(x) + C$ for all $x \in (a, b)$ and some constant $C \in \mathbb{R}$.*

Definition 3.12 (Monotonic, Increasing, Decreasing)
The function f defined on the interval (a, b) is said to be

1. *increasing* in (a, b) if for all $x_1, x_2 \in (a, b)$:
   $x_1 < x_2 \Rightarrow f(x_1) < f(x_2)$.
2. *non-decreasing* in (a, b) if for all $x_1, x_2 \in (a, b)$:
   $x_1 < x_2 \Rightarrow f(x_1) \leq f(x_2)$
3. *decreasing* in (a, b) if for all $x_1, x_2 \in (a, b)$:
   $x_1 < x_2 \Rightarrow f(x_1) > f(x_2)$
4. *non-increasing* in (a, b) if for all $x_1, x_2 \in (a, b)$:
   $x_1 < x_2 \Rightarrow f(x_1) \geq f(x_2)$
5. *monotonic* in (a, b) if f is either increasing or non-decreasing or decreasing or non-increasing in (a, b).

Theorem 3.20 (Characterization of Monotonicity)
If f is differentiable on the interval (a, b), then

1. *If $f'(x) > 0$ for all $x \in (a, b)$, then f is increasing on (a, b);*
2. *If $f'(x) < 0$ for all $x \in (a, b)$, then f is decreasing on (a, b);*
3. *If $f'(x) = 0$ for all $x \in (a, b)$, then f is constant on (a, b);*

Definition 3.13 (Concavity, Concave up/down)
If the function f is differentiable on the interval (a, b), then we say that f is

1. *concave upward* on (a, b) if f' is increasing on (a, b).
2. *concave downward* on (a, b) if f' is decreasing on (a, b).

Figure 3.3: *Pierre de Fermat (1601–1665)*

Theorem 3.21 (Characterization of Concavity)
Suppose the function f is twice differentiable on the interval (a, b). Then

1. *If $f''(x) > 0$ for all $x \in (a, b)$, then f is concave upward on (a, b).*
2. *If $f''(x) < 0$ for all $x \in (a, b)$, then f is concave downward on (a, b).*

Theorem 3.22 (Fermat's Theorem)
If f has a local extremum at c then $f'(c) = 0$ or $f'(c)$ does not exist.

Definition 3.14 (Critical Numbers)
Given a function f with domain D_f. The point $c \in D_f$ is called a *critical number* of f if either $f'(c) = 0$ or $f'(c)$ does not exist.

Theorem 3.23 (First Derivative Test)
Suppose c is a critical point of f and f is continuous at c. If there exists a $\delta > 0$ such that:

1. *$f'(x) > 0$ for all $x \in (c - \delta, c)$ and $f'(x) < 0$ for all $x \in (c, c + \delta)$, then $f(c)$ is a relative maximum.*
2. *$f'(x) < 0$ for all $x \in (c - \delta, c)$ and $f'(x) > 0$ for all $x \in (c, c + \delta)$, then $f(c)$ is a relative minimum*
3. *If f keeps constant sign on $(c - \delta, c) \cup (c, c + \delta)$ then $f(c)$ is not an extreme value.*

Theorem 3.24 (First Derivative Test for Absolute Extrema)
Suppose c is a critical point of f defined on an interval, and suppose f is continuous at c. If

1. *$f'(x) > 0$ for all $x < c$ and $f'(x) < 0$ for all $x > c$, then $f(c)$ is an absolute maximum of f.*
2. *$f'(x) < 0$ for all $x < c$ and $f'(x) > 0$ for all $x > c$, then $f(c)$ is an absolute minimum of f.*

Theorem 3.25 (Second Derivative Test)
Suppose that $f'(c) = 0$ and that $f''(c)$ exists. Then

1. *If $f''(c) > 0$ then $f(c)$ is a local minimum.*
2. *If $f''(c) < 0$ then $f(c)$ is a local maximum.*

3.2 Exercises

3.2.1 Two-Sided Limits

Find the following limits if they exist:

1. $\lim_{x \to 3} \frac{2x-6}{x-3}$ has

2. $\lim_{x \to 3} \frac{x-3}{x^2-6x-9}$ has

3. $\lim_{x \to 1} \frac{x-1}{x^2-5x+4}$ has

4. $\lim_{x \to 0} \frac{1-\frac{1}{x}}{1+\frac{1}{x}}$ has

5. $\lim_{x \to 1} \frac{3x-3}{\sqrt{x}-1}$ has

6. $\lim_{x \to 2} \frac{\frac{1}{\sqrt{2}}-\frac{1}{\sqrt{x}}}{x-2}$ has

7. $\lim_{x \to 1} \frac{x^{2/3}-1}{x-1}$ has

8. $\lim_{x \to 0} \frac{\tan^2 x}{x}$ has

9. $\lim_{x \to \pi} \frac{\sin x}{x-\pi}$ has

10. $\lim_{x \to 0} \frac{\sin 2x}{\sin 3x}$ has

11. $\lim_{x \to \pi/4} \frac{\tan(\pi-4x)}{x-\pi/4}$ has

12. $\lim_{x \to 0} |x| \cos x$ has

13. $\lim_{x \to \pi/4} \frac{\sin(x+\frac{1}{4}\pi)-1}{x-\frac{1}{4}\pi}$ a

3.2.2 One-Sided Limits

Find the following limits if they exist:

1. $\lim_{x\to 8^+} \frac{|8-x|}{8-x}$ a

2. $\lim_{x\to 0^+} \frac{\sin x}{\sqrt{x}}$ has

3. $\lim_{x\to 4^+} \frac{1}{x-4}$ a

4. $\lim_{x\to 4^-} \frac{1}{x-4}$ a

5. $\lim_{x\to 0^+} x \sin\frac{1}{x}$ a

6. $\lim_{x\to 0} x \sin\frac{1}{x}$ a

3.2.3 Limits at Infinity

Find the following limits if they exist:

1. $\lim_{x\to\infty} \frac{x+1}{x-3}$ a

2. $\lim_{x\to\infty} \frac{x^2+1}{2x^2+4}$ has

3. $\lim_{x\to\infty} \frac{1-x^3}{x^3+4x+3}$ a

4. $\lim_{x\to\infty} \frac{x+1}{x^2-3}$ a

5. $\lim_{x\to\infty} \frac{5x^3+4(\sqrt{x})^5+2x+\tanh x}{4x^3+2x+4}$ a

6. $\lim_{x\to\infty} \left(\sqrt{x^2+4x}-\sqrt{x}\right)$ a

7. $\lim_{x\to\infty} \left(\sqrt{x^2+4x}-x\right)$ a

8. $\lim_{x\to\infty} \frac{1}{1+e^{-x}}$ a

9. $\lim_{x\to-\infty} \frac{1}{1+e^{-x}}$ a

3.2.4 Squeeze Play

Find the following limits if they exist.

1. $\lim_{x\to\infty} \frac{\sin x}{x}$ a

2. $\lim_{x\to 0} x \sin\frac{1}{x}$ a

3. $\lim_{n\to\infty} \frac{e^n}{n!}$ has

4. $\lim_{x\to -3} (x+3)^2 \sin\frac{1}{x+3}$ a

5. $\lim_{x\to\pi} (x-\pi)\cos^2\frac{1}{x-\pi}$ a

6. $\lim_{x \to 1} |x-1| \sin x$ a

3.2.5 L'Hôpital's Rule

Find the following limits if they exist:

1. $\lim_{x \to \infty} \frac{3^x}{\ln x}$ has
2. $\lim_{x \to \infty} \left(1 + \frac{a}{x}\right)^x$ has
3. $\lim_{x \to 0} \frac{\tan \pi x}{e^x - 1}$ a
4. $\lim_{x \to 0+} \frac{\ln x}{\cot x}$ a
5. $\lim_{x \to 0} \frac{-6x}{\tan 4x}$ a
6. $\lim_{x \to 0} \left(\frac{1}{\sin x} - \frac{1}{x}\right)$ a
7. $\lim_{x \to \infty} \left(\sqrt{x^2 + 2x} - x\right)$ a
8. $\lim_{x \to \infty} \left(\cos \frac{1}{x}\right)^x$ a
9. $\lim_{x \to \infty} \left(x^3 + 1\right)^{1/\ln x}$ a
10. $\lim_{x \to 0} |\sin x|^x$ a
11. $\lim_{x \to \pi/2} \frac{\ln(\sin x)}{x - \frac{\pi}{2}}$ a
12. $\lim_{x \to \infty} \frac{1}{x} \int_0^x \sin\left(\frac{1}{1+t}\right) dt$ a
13. $\lim_{x \to 2} \frac{-\arctan(4x - 8)}{\arcsin(7x - 14)}$ a

3.2.6 Continuity, Removable Dicontinuities, etc.

1. Sketch the graph of the function

$$f(x) = \begin{cases} 1 & x \leq 0 \\ x^2 & 0 < x < 1 \\ 1 & 1 \leq x < 2 \\ x & 2 \leq x \end{cases}$$

and classify the discontinuities (if any) as being a removable discontinuity, a jump discontinuity, or an infinite discontinuity. has

2. Sketch the graph of the function

$$f(x) = \begin{cases} x+7 & x < -3 \\ |x-2| & -3 < x < -1 \\ x^2 - 2x & -1 < x < 3 \\ 2x - 3 & 3 \leq x \end{cases}$$

and classify the discontinuities (if any) as being a removable discontinuity, a jump discontinuity, or an infinite discontinuity. has

3. Find where the function

$$f(x) = \frac{x+1}{x^2 + 5x + 4}$$

is continuous. has

4. Find where the function

$$f(x) = \frac{x+3}{x-3}$$

is continuous and classify any points of discontinuity. has

5. Let

$$f(x) = \frac{x-2}{x^2-4}, \quad x \in \mathbb{R} \setminus \{-2, 2\}.$$

Can f be defined at $x = 2$ such that f is continuous at $x = 2$? has

6. Let

$$f(x) = \frac{x-2}{x^2+3x-10}, \quad x \in \mathbb{R} \setminus \{-5, 2\}.$$

Can f be defined at 2 such that f is continuous at 2? ha

7. Let

$$f(x) = \frac{x-3}{x^2+8x+12}, \quad x \in \mathbb{R} \setminus \{-2, -6\}$$

.Can f be defined at -6 such that f is continuous at -6? ha

3.2.7 Intermediate Value Theorem

Recall and interpret the Intermediate Value Theorem. The real valued, continuous function f is defined on the closed and bounded interval $[a, b]$. Which of the following must be true?

1. If either $f(a)$ or $f(b)$ is larger than c, then there exists $x \in [a, b]$ such that $f(x) = c$. a

2. The function has either a maximum or minimum value but we cannot say which. a

3. If either $f(a)$ or $f(b)$ is smaller than c, then there exits $x \in [a, b]$ such that $f(x) = c$. a

4. If $f(a)$ is smaller than c and $f(b)$ is larger than c, then there exists $x \in (a, b)$ such that $f(x) = c$. a

5. If $f(a)$ and $f(b)$ are both larger than c, then there exists $x \in [a, b]$ such that $f(x) = c$. a

6. The function has a derivative on the open interval (a, b). a

7. If there exists $x \in [a, b]$ such that $f(x) = c$, then either $f(a) < c < f(b)$ or $f(b) < c < f(a)$. a

8. None or the above. a

3.2.8 Mean Value Theorem

Select the correct formulation for the Mean Value Theorem:

1. Let f be continuous on $[a, b]$ and suppose f' is continuous on (a, b). Then there exists $c \in [a, b]$ such $f'(c)\big(f(b) - f(a)\big) = b - a$. a

2. If f is continuous on $[a, b]$ and f' exists on (a, b), then there exists a point $c \in (a, b)$ such that $f'(c)(b - a) = f(b) - f(a)$. a

3. If f is continuous on $[a, b]$, then f is differentiable on (a, b) and $f'(c)(b - a) = f(b) - f(a)$. a

4. If f is differentiable on (a, b), then f is continuous on $[a, b]$ and $f'(c)(b - a) = f(b) - f(a)$. a

5. If $f'(x) = 0$ for all $x \in (a, b)$, then f equals a constant. a

6. Let f be differentiable on (a, b). Then $f'(c)(b-a) = f(b) - f(a)$ for all $c \in (a, b)$. a

7. Let f be differentiable on (a, b) and continuous on $[a, b]$. Then for all $c \in (a, b)$, we have $f'(c)(b-a) = f(b) - f(a)$. a

8. If for some $c \in (a, b)$ we have $f'(c)(b-a) = f(b)-f(a)$, then it follows f is differentiable on (a, b) and continuous on $[a, b]$. a

9. For some $c \in (a, b)$ it follows $f'(c)(b-a) = f(a) - f(b)$ provided f is continuous on $[a, b]$ and differentiable on (a, b). a

10. None of the above. a

3.2.9 Extreme Value Theorem

The real valued, continuous function f is defined on the closed and bounded interval $[a, b]$. Which of the following must be true?

1. If $f(a)$ and $f(b)$ are both larger than c, then there exists $x \in [a, b]$ such that $f(x) = c$. a

2. The function has a derivative on the open interval (a, b). a

3. There exists $x \in [a, b]$ such that $f(x) \geq f(y)$ for all $y \in [a, b]$. a

4. If either $f(a)$ or $f(b)$ is larger than c, then there exists $x \in [a, b]$ such that $f(x) = c$. a

5. The function has either a maximum or minimum value but we cannot say which. a

6. There exists $x \in [a, b]$ such that $f(x) = f(y)$ for all $y \in [a, b]$. a

7. If either $f(a)$ or $f(b)$ is smaller than c, then there exists $x \in [a, b]$ such that $f(x) = c$. a

8. If there exists $x \in [a, b]$ such that $f(x) = c$, then either $f(a) < c < f(b)$ or $f(b) < c < f(a)$. a

9. None of the above. a

3.2.10 Equations of Tangents and Normals

Find the equation for the tangent and normal to the given curve at the given point:

1. $f(x) = \sin x, \quad (\frac{\pi}{6}, \frac{1}{2})$ has

2. $f(x) = \sqrt{x+1}, \quad (3, 2)$ has

3. $f(x) = 1 - e^{-x}, \quad (0, 0)$ has

4. $f(x) = 1 + \ln x, \quad (3, 1 + \ln 3)$ has

3.2.11 The Derivative

1. Recall the Definition of a Derivative a

2. Use the Definition to compute the Derivative of the following Functions at x_0:

   (a) $f(x) := \dfrac{x}{x+1}, \quad x_0 = 2$ ha

   (b) $f(x) = \dfrac{1}{\sqrt{x}}, \quad x_0 = 4$ ha

   (c) $f(x) = \sin x, \quad x_0 = 0$ a

   (d) $f(x) = \begin{cases} x^2 \sin(\frac{1}{x}) + x(1-x) & x \neq 0 \\ 0 & \text{else.} \end{cases}, \quad x_0 = 0.$ a

3. Determine whether each of the following function is (α) differentiable (β) continuous at x_0:

   (a) $f(x) = \begin{cases} x^2 & x < 1 \\ 2x - 1 & x \geq 1 \end{cases},$

   at $x_0 = 1.$ ha

   (b) $f(x) = \begin{cases} x^2 + 1 & x \leq 1 \\ -x^2 + 2x + 1 & x > 1 \end{cases},$

   at $x_0 = 1.$ a

   (c) $f(x) = \begin{cases} 1 - x^2 & x < 2 \\ -3 & x = 2 \\ x^2 - 8x + 9 & x > 2 \end{cases},$

   at $x_0 = 2.$ a

3.2.12 Applying the Differentation Rules

Rational Functions

1. $f(x) = (x - 2x^2)^2$ ha

2. $f(x) = \dfrac{1}{x\sqrt{3+2x^2}}$ ha

3. $f(x) = \dfrac{x^2-1}{x^3+1}$ a

4. $f(x) = \left(\left(x^{-1}+2x^{-2}\right)^3 + 3x^{-3}\right)^4$ a

5. $g(x) = \dfrac{x^2+x+1}{x^2-x+1}$ a

6. $g(t) = \dfrac{t^2+5t+4}{t^2+t-20}$ a

7. $f(x) = \dfrac{x^3-1}{x^4+1}(x^2+1)$ a

8. $f(x) = \left(1-\dfrac{1}{x}\right)\left(2-\dfrac{1}{x}\right)$ a

9. $f(x) = \left(x^2 - \dfrac{1}{x^3+1}\right)^5$ a

10. $f(x) = \left(\dfrac{4x+3}{5x-2}\right)^3$ a

Trigonometric Functions

11. $f(x) = \sin x \cos x$ a

12. $f(x) = x^2 \cos x$ a

13. $f(x) = \dfrac{\sec x}{\tan^2 x + 1}$ ha

14. $f(x) = \sqrt[3]{\sec^2(\tan(x))}$ ha

15. $f(x) = \tan(\sec(x+1))$ a

16. $f(x) = 2x^2 \tan x \sec x$ a

17. $f(x) = \cos(\sin x)$ a

18. $f(x) = (1 + \sin x)^{\frac{1}{3}}$ a

19. $f(x) = \left(\sec(\tan x)\right)^{\frac{1}{4}}$ a

20. $f(x) = \cot^2\left(2\sqrt{3x+1}\right)$ a

21. $f(x) = x^2 \sec\left(x^{-3}\right)$ a

22. $f(x) = \cos^3(2x)\tan(4x)$ a

Exponential and Logarithmic Functions

23. $f(x) = 2^{5x}3^{\ln x}$ a

24. $f(x) = \ln\left(\ln(2x+1)\right)$ a

25. $f(x) = (x+1)^x$ ha

26. $f(x) = x^{x^x}$ a

Miscellaneous

27. $f(x) = \cos(2^x + 2^{-x})$ a

28. $f(x) = (\tan x)^{\ln(x^x)}$ a

29. $f(x) = (\sin x)^{\cos x}$ a

30. $f(x) = \ln|\sec x + \tan x|$ a

31. $f(x) = \tan\sqrt{e^{-3x}}$ a

32. $f(x) = \sec\left(e^{\tan(4x^2)}\right)$ a

33. $f(x) = e^{4\ln x}$ a

34. $f(x) = e^{\sin(2x)}\ln\left(\cos e^{2x}\right)$ a

35. $f(x) = \sin\left(\operatorname{arcsec}(\ln x)\right)$ a

36. $f(x) = \frac{1}{3}\arcsin(3x - 4x^2)$ a

37. $f(x) = \cos(\operatorname{arcsec}(\ln x))$ a

38. $f(x) = e^{\operatorname{arccot} x}$ a

39. $f(x) = \arctan\left(\dfrac{x}{1+x^2}\right)$ a

3.2.13 Implicit Differentiation

1. Find dy/dx in terms of x and y:

   (a) $\tan\sqrt{x+y} = 2x$ a

   (b) $|x| + |y| = 1$ a

   (c) $x^{2/3} + y^{2/3} = 1$ a

   (d) $\tan(xy) = x$ a

   (e) $x^2 - 3xy + 4y^2x^2 = 18x^2$ a

2. Find Equations for the Tangent and Normal Lines at the indicated Point, respectively:

   (a) $9x^2 + 4y^2 = 72, \quad (2,3)$ a

   (b) $y^2 + 2xy + x^2 = 9, \quad (2,1)$ a

   (c) $x = \cos y, \quad (\frac{1}{2}, \frac{\pi}{3})$ a

3.2.14 Extreme Values

Find the critical numbers and classify them as relative and absolute:

1. $f(x) = x^2 - 4x + 1, \quad x \in [0,3]$ has

2. $f(x) = x^4 - x^2 + 4$ has

3. $f(x) = x^3(4-x)^2$ has

4. $f(x) = \dfrac{x^2}{1+x^2}, \quad x \in [-2,6]$ ha

5. $f(x) = 6 + 216x^2 + \frac{2}{x}, \quad x \in (0,\infty)$ ha

6. $f(x) = \dfrac{x}{4+x^2}, \quad x \in [-3,1]$ ha

7. $f(x) = x\sqrt{4 - x^2}$ has

8. $f(x) = 2\cos^3 x + 3\sin x, \quad x \in [0, \pi]$ has

9. $f(x) = \begin{cases} |x+1| & -3 \le x < 0 \\ x^2 - 4x + 2 & 0 \le x < 3 \\ 2x - 7 & 3 \le x < 4 \end{cases}$ has

3.2.15 Concavity, Inflection Points

Decribe the concavity of the graph of f and find the points of inflections:

1. $f(x) = x^3 - 3x + 2$ ha

2. $f(x) = \dfrac{6x}{x^2 + 1}$ ha

3. $f(x) = 2x^3 + 3x^2$ ha

4. $f(x) = x\sqrt{4 - x^2}$ ha

5. $f(x) = x + \sin x$ has

6. $f(x) = \begin{cases} x^3 & x < 1 \\ 3x - 2 & x \ge 1 \end{cases}$ has

3.2.16 Monotonicity

Find the intervals on which f increases/decreases:

1. $f(x) = x + \frac{1}{x}$ ha

2. $f(x) = 3x^3 + 7x^2 + 7$ ha

3. $f(x) = \dfrac{x^2 + 1}{x^2 - 1}$ ha

4. $f(x) = x + \sin x$ ha

3.2.17 Graph Sketching

Sketch the graph of the function f:

1. $f(x) = x^3 - 2x^2 + x + 1$ ha

2. $f(x) = \dfrac{x - 4}{x^2}$ ha

3. $f(x) = \dfrac{2x^2}{x+1}$ ha

4. $f(x) = 3 - |x^2 - 1|$ ha

5. $f(x) = x(x-1)^{\frac{1}{5}}$ ha

3.2.18 Maximum/Minimum Problems

1. Find the dimensions of the isosceles triangle of largest area with perimeter 12. a

2. Find the point on the parabola $y = x^2$ which is closest to the point $(0, 3)$. a

3. A rectangular garden 12 square feet in area is to be fenced off against rabbits. Find the least possible length of fencing if one side of the garden is already protected by a barn. a

4. A rectangular playground is to be enclosed by a fence and divided in 4 pieces by 3 fences parallel to one side of the playground. 1060 feet of fencing is used. Find the dimension of the playground which will have the largest total area. a

5. What is the maximum possible area for a triangle inscribed in a circle of radius r. a

6. Conical paper cups are usually made so that the depth is $\sqrt{2}$ times the radius of the rim. Show that this design requires the least amount of paper per unit volume. has

7. What is the maximum volume for a rectangular box (square base, no top) made from 12 square feet of cardboard? a

3.3 Hints to the Exercises

3.3.1 Two-Sided Limits

Find the following limits if they exist:

1. First reduce the argument of the limit operator.

2. First check whether the denominator polynomial is 0 for $x = 3$. If not, apply Theorem 3.2:4.

3. First reduce the argument of the limit operator. Then check, whether Theorem 3.2:4 can be applied.

4. Expand the fraction by x.

5. Reduce the quotient!

6. Reduce the fraction using

$$x - 2 = \left(\sqrt{x} - \sqrt{2}\right)\left(\sqrt{x} + \sqrt{2}\right).$$

7. Factor the numerator and denominator of the fraction and reduce it.

8. Write tan in terms of sin and cos and use Theorem 3.3:2, 3 and 6.

9. Use the substitution $y = x - \pi$, the addition formula for sin (Theorem 2.4:1) and Theorem 3.3:6.

10. Expand the argument of the limit operator by $2x$ and use Theorem 3.3:6.

11. Use L'Hôpital's Rule (Theorem 3.8) or Theorem 3.3:6.

12. Use the inequality $-1 \leq \cos x \leq 1$ for $x \in \mathbb{R}$ in conjunction with the "Squeezing Theorem" (Theorem 3.7).

3.3.2 One-Sided Limits

Find the following limits if they exist:

2. Expand the argument of the limit by $\sqrt{x}$ and apply Theorem 3.3:6,1.

3.3.3 Limits at Infinity

Find the following limits if they exist:

2. Factor out x^2 in the numerator and denominator of the argument of the limit operator and reduce. Then apply Theorem 3.4 and Theorem 3.3:8.

3.3.4 Squeeze Play

Find the following limits if they exist.

3. Establish the inequality

$$0 \le \frac{e^n}{n!} \le \frac{e^3}{2}\frac{1}{n}$$

for all $n \in \mathbb{N}$ with $n \ge 4$ and apply the "Squeezing Theorem" (Theorem 3.7).

3.3.5 L'Hôpital's Rule

Find the following limits if they exist:

1. Apply L'Hôpital's Rule and simplify the limit.

2. Rewrite

$$\left(1 + \frac{a}{x}\right)^x = e^{x \ln(1+\frac{a}{x})}$$

and use the Substitution Rule (Theorem 3.5) and L'Hôpital's Rule (Theorem 3.8).

3.3.6 Continuity, Removable Dicontinuities, etc.

1. Check half-sided limits at 0, 1 and 2!

2. Check the half-sided limits at -3, -1 and 3!

3. Determine the domain of f. Also check for common factors in the numerator and denominator of the rational term defining f.

4. Determine the domain. Check the half-sided limits at the zeroes of the denominator!

5. Look for common divisiors!

6. Look for common divisiors!

7. Look for common divisiors!

3.3.7 Intermediate Value Theorem

Recall and interpret the Intermediate Value Theorem. The real valued, continuous function f is defined on the closed and bounded interval $[a, b]$. Which of the following must be true?

3.3.8 Mean Value Theorem

Select the correct formulation for the Mean Value Theorem:

3.3.9 Extreme Value Theorem

The real valued, continuous function f is defined on the closed and bounded interval $[a, b]$. Which of the following must be true?

3.3.10 Equations of Tangents and Normals

Find the equation for the tangent and normal to the given curve at the given point:

1. Compute the slope m of the graph of f at the point $(\frac{\pi}{6}, \frac{1}{2})$. Use the point-slope form of the line equation. The slope $m^{\perp}$ of the normal line is $-1/m$.
2. Compute the slope m of the graph of f at the point $(3, 2)$. Use the point-slope form of the line equation. The slope $m^{\perp}$ of the normal line is $-1/m$
3. Compute the slope m of the graph of f at the point $(0, 0)$. Use the point-slope form of the line equation. The slope $m^{\perp}$ of the normal line is $-1/m$
4. Compute the slope m of the graph of f at the point $(3, \ln 3))$. Use the point-slope form of the line equation. The slope $m^{\perp}$ of the normal line is $-1/m$

3.3.11 The Derivative

2. Use the Definition to compute the Derivative of the following Functions at x_0:

   (a) Try to factor out $x - 2$ in the numerator of the argument of the differential quotient!

   (b) Try to factor out $x - 4$

5. Determine whether each of the following function is (α) differentiable (β) continuous at x_0:

   (a) investigate the left- and right-sided limits at $x_0 = 1$

3.3.12 Applying the Differentation Rules

Rational Functions

1. Apply the Chain Rule!
2. Try the Quotient Rule!

Trigonometric Functions

13. Can you simplify the term before you differentiate?

14. Simplify the term before you differentiate!

Exponential and Logarithmic Functions

25. You can use "logarithmic differentiation" or rewrite the expression using the identity $a^b = e^{b \ln a}$

Miscellaneous

3.3.13 Implicit Differentiation

1. Find dy/dx in terms of x and y:

7. Find Equations for the Tangent and Normal Lines at the indicated Point, respectively:

3.3.14 Extreme Values

Find the critical numbers and classify them as relative and absolute:

1. Determine the zeroes of f' and apply Fermat's Theorem.

2. Determine the zeroes of f' and apply Fermat's Theorem. Apply the First Derivative Test. Does the Extreme Value Theorem apply?

3. Determine the zeroes of f', apply Fermat's Theorem and the First Derivative Test. Does the Extreme Value Theorem apply?

4. Find the zeroes of f', apply Fermat's Theorem and the First Derivative Test.

5. Find the zeroes of f' on the interval $(0, \infty)$ and apply the First Derivative Test.

6. Determine the zeroes of f', apply Fermat's Theorem and the First Derivative Test. Does the Extreme Value Theorem apply?

7. What is the domain D_f of f? Determine f'. Is f' defined for all $x \in D_f$? Apply Fermat's Theorem.

8. Determine f' and find its zeroes on $(0, \pi)$. Are there any points in $(0, \pi)$ where f is not differentiable?

9. Where is f differentiable on $[-3, 4)$? For which $x \in (-3, 4)$ is $f'(x) = 0$?

3.3.15 Concavity, Inflection Points

Decribe the concavity of the graph of f and find the points of inflections:

1. Consider the second derivative of f and check where f'' is positive and where it is negative!
2. Consider the second derivative of f and check where f'' is positive and negative!
3. Determine f'' and find the intervals where f'' is positive and where it is negative.
4. Determine f'' and find the intervals where f'' is positive and where it is negative.
5. Determine f'' and find the intervals where f'' is positive and where it is negative.
6. Determine f'' and find the intervals where f'' is positive and where it is negative.

3.3.16 Monotonicity

Find the intervals on which f increases/decreases:

1. Determine the intervals on which f' is positive and where it is negative.
2. Determine the intervals on which f' is positive and where it is negative.
3. Determine the intervals on which f' is positive and where it is negative.
4.

3.3.17 Graph Sketching

Sketch the graph of the function f:

1. Determine (i) Domain, (ii) Symmetry, (iii) x-Intercepts, (iv) y-Intercepts, (v) Monotonicty, (vi) Extrema, (vii) Concavity, (viii) Inflection Points, (ix) Asymptotes
2. Determine (i) Domain, (ii) Symmetry, (iii) x-Intercepts, (iv) y-Intercepts, (v) Monotonicty, (vi) Extrema, (vii) Concavity, (viii) Inflection Points, (ix) Asymptotes
3. Determine (i) Domain, (ii) Symmetry, (iii) x-Intercepts, (iv) y-Intercepts, (v) Monotonicty, (vi) Extrema, (vii) Concavity, (viii) Inflection Points, (ix) Asymptotes

4. Determine (i) Domain, (ii) Symmetry, (iii) x-Intercepts, (iv) y-Intercepts, (v) Monotonicty, (vi) Extrema, (vii) Concavity, (viii) Inflection Points, (ix) Asymptotes

5. Determine (i) Domain, (ii) Symmetry, (iii) x-Intercepts, (iv) y-Intercepts, (v) Monotonicty, (vi) Extrema, (vii) Concavity, (viii) Inflection Points, (ix) Asymptotes

3.3.18 Maximum/Minimum Problems

6. Show that the surface area of the conical cup has its absolute minimum if $h = \sqrt{2}r$!

3.4 Answers to the Exercises

3.4.1 Two-Sided Limits

Find the following limits if they exist:

1. 2
2. 0
3. $-\frac{1}{3}$
4. -1
5. 0
6. $\frac{1}{8}\sqrt{2}$
7. $\frac{2}{3}$
8. 0
9. -1
10. $\frac{2}{3}$
11. -4
12. 0
13. 0

3.4.2 One-Sided Limits

Find the following limits if they exist:

1. -1
2. 0
3. ∞
4. -∞
5. 0
6. 0

3.4.3 Limits at Infinity

Find the following limits if they exist:

1. 1
2. $\frac{1}{2}$
3. -1
4. 0
5. $\frac{5}{4}$
6. ∞
7. 2
8. 1
9. 0

3.4.4 Squeeze Play

Find the following limits if they exist.

1. 0
2. 0
3. 0
4. 0
5. 0
6. 0

3.4.5 L'Hôpital's Rule

Find the following limits if they exist:

1. ∞
2. e^a
3. π
4. 0
5. -$\frac{3}{2}$
6. 0

7. 1
8. 1
9. e^3
10. 1
11. 0
12. 0
13. -$\frac{4}{7}$

3.4.6 Continuity, Removable Dicontinuities, etc.

1. f has jump discontinuities at $x = 0$ and $x = 2$.
2. f has a jump discontinuity at $x = -3$ and a removable discontinuity at $x = -1$. f is continuous at $x = 1$.
3. f is continuous on $\mathbb{R} \setminus \{-1, -4\}$, but has a removable discontinuity at $x = -1$.
4. f is continuous on $D_f := \mathbb{R} \setminus \{3\}$ and has an infinite discontinuity at $x = 3$.
5. Yes, define $f(2) := 1/4$.
6. Yes, define $f(2) := 1/7$.
7. No, f has an infinite discontinuity at $x = -6$.

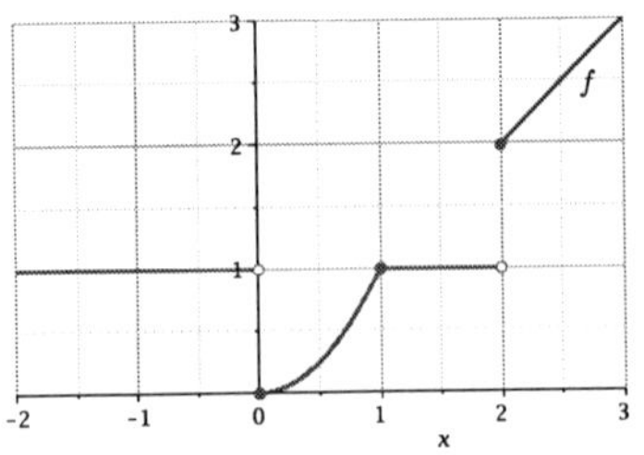

Figure 3.4:

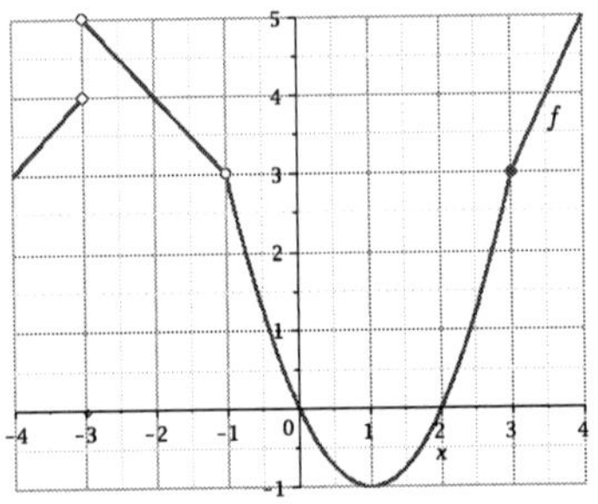

Figure 3.5:

3.4.7 Intermediate Value Theorem

Recall and interpret the Intermediate Value Theorem. The real valued, continuous function f is defined on the closed and bounded interval $[a, b]$. Which of the following must be true?

1. true
2. false
3. true
4. true
5. false
6. false
7. false
8. false

3.4.8 Mean Value Theorem

Select the correct formulation for the Mean Value Theorem:

1. false
2. true
3. false
4. false
5. true
6. false
7. false
8. false
9. true
10. false

3.4.9 Extreme Value Theorem

The real valued, continuous function f is defined on the closed and bounded interval $[a, b]$. Which of the following must be true?

1. false	4. true	7. true
2. false	5. false	8. false
3. true	6. false	9. false

3.4.10 Equations of Tangents and Normals

Find the equation for the tangent and normal to the given curve at the given point:

1. Tangent line: $T(x) = \left(\frac{1}{2}\sqrt{3}\right)x + \left(\frac{1}{2} - \frac{\pi}{12}\sqrt{3}\right)$
   Normal line: $N(x) = \left(-\frac{2}{3}\sqrt{3}\right)x + \left(\frac{1}{2} + \frac{\pi}{9}\sqrt{3}\right)$.

2. Tangent line: $T(x) = \frac{1}{4}x + \frac{5}{4}$
   Normal line: $N(x) = -4x + 14$.

3. Tangent line: $T(x) = x$
   Normal line: $N(x) = -x$.

4. Tangent line: $T(x) = \frac{1}{3}x + \ln 3$
   Normal line: $N(x) = -3x + (10 + \ln 3)$.

3.4.11 The Derivative

1. We say that f is *differentiable* at c if the limit

$$\lim_{x \to c} \frac{f(x) - f(c)}{x - c}$$

   exists. If the limit exists, it is called the *derivative* of f at c and denoted by $f'(c)$.

2. Use the Definition to compute the Derivative of the following Functions at x_0:

   (a) $\frac{1}{9}$
   (b) $-\frac{1}{16}$
   (c) 1
   (d) 1

3. Determine whether each of the following function is (α) differentiable (β) continuous at x_0:

   (a) (α) differentiable (β) continuous
   (b) (α) not differentiable (β) continuous
   (c) (α) differentiable (β) continuous

3.4.12 Applying the Differentation Rules

Rational Functions

1. $16x^3 - 12x^2 + 2x$
2. $-\frac{3+4x^2}{x^2\left(3+2x^2\right)^{3/2}}$
3. $-\frac{x(x-2)}{\left(x^2-x+1\right)^2}$
4. $-\dfrac{384\left(2x^3+3x^2+6x+4\right)^3\left(x^3+2x^2+5x+4\right)}{x^{25}}$
5. $-\dfrac{2\left(x^2-1\right)}{\left(x^2-x+1\right)^2}$
6. $-\dfrac{4\left(t^2+12t+26\right)}{\left(t^2+t-20\right)^2}$
7. $\dfrac{x\left(x^7+5x^3-x^5+3x+2x^4+4x^2-2\right)}{\left(x^4+1\right)^2}$
8. $\dfrac{3x-2}{x^3}$
9. $\dfrac{5x\left(x^5+x^2-1\right)^4\left(2x^6+4x^3+3x+2\right)}{\left(x^3+1\right)^6}$
10. $-\dfrac{69\left(4x+3\right)^2}{\left(5x-2\right)^4}$

Trigonometric Functions

11. $\cos^2(x) - \sin^2(x) = \cos(2x)$
12. $2x\cos(x) - x^2\sin(x)$
13. $-\sin x$
14. $\frac{2}{3}\left(\sec\left(\tan(x)\right)\right)^{2/3}\tan\left(\tan(x)\right)\sec^2(x)$
15. $\sec^2\left(\sec(x+1)\right)\sec(x+1)\tan(x+1)$
16. $4x\tan(x)\sec(x) + 2x^2\sec^3(x) + 2x^2\tan^2(x)\sec(x)$
17. $-\sin\left(\sin(x)\right)\cos(x)$

18. $\dfrac{\cos(x)}{3(1+\sin(x))^{2/3}}$

19. $\frac{1}{4}\Big(\sec\big(\tan(x)\big)\Big)^{1/4}\tan\big(\tan(x)\big)\sec^2(x)$

20. $\dfrac{-6\cot\big(2\sqrt{3x+1}\big)\csc^2\big(2\sqrt{3x+1}\big)}{\sqrt{3x+1}}$

21. $\dfrac{2x^3\sec\big(x^{-3}\big)-3\sec\big(x^{-3}\big)\tan\big(x^{-3}\big)}{x^2}$

22. $-6\cos^2(2x)\sin(2x)\tan(4x)+4\cos^3(2x)\sec^2(4x)$

Exponential and Logarithmic Functions

23. $\dfrac{(5x\ln 2)2^{5x}3^{\ln x}+(\ln 3)2^{5x}3^{\ln x}\ln x}{x}$

24. $\dfrac{2}{(2x+1)\ln(2x+1)}$

25. $(x+1)^x\left[\ln(x+1)+\dfrac{x}{x+1}\right]$

26. $x^{x^x+x-1}\big[x\ln x(\ln x+1)+1\big]$

Miscellaneous

27. $(-\ln 2)\left[2^x-2^{-x}\right]\sin\big(2^x+2^{-x}\big)$

28. $(\tan x)^{\ln x^x}\left[\ln x\ln(\tan x)+\ln(\tan x)+x\ln\left(\dfrac{\sec^2 x}{\tan x}\right)\right]$

29. $(\sin x)^{\cos x}\left[-\sin x\ln(\sin x)+\dfrac{\cos^2 x}{\sin x}\right]$

30. $\dfrac{\sec(x)\tan(x)+\sec^2 x}{|\sec(x)+\tan(x)|}$

31. $-\frac{3}{2}e^{-\frac{3}{2}x}\sec^2\big(e^{-\frac{3}{2}x}\big)$

32. $8xe^{\tan(4x^2)}\sec^2(4x^2)\sec\big(e^{\tan(4x^2)}\big)\tan\big(e^{\tan(4x^2)}\big)$

33. $\dfrac{4e^{4\ln(x)}}{x}$

34. $2e^{\sin 2x}\left[\cos 2x\ln\big(\cos(e^{2x})\big)-e^{2x}\tan(e^{2x})\right]$

35. $\dfrac{1}{x\ln^2 x\sqrt{\ln^2 x - 1}}$

36. $\dfrac{3-8x}{3\sqrt{1-9x^2+24x^3-16x^4}}$

37. $-\dfrac{1}{x\ln^2 x}$

38. $-\dfrac{e^{\operatorname{arccot}(x)}}{x^2+1}$

39. $\dfrac{1-x^2}{x^2+(x^2+1)^2}$

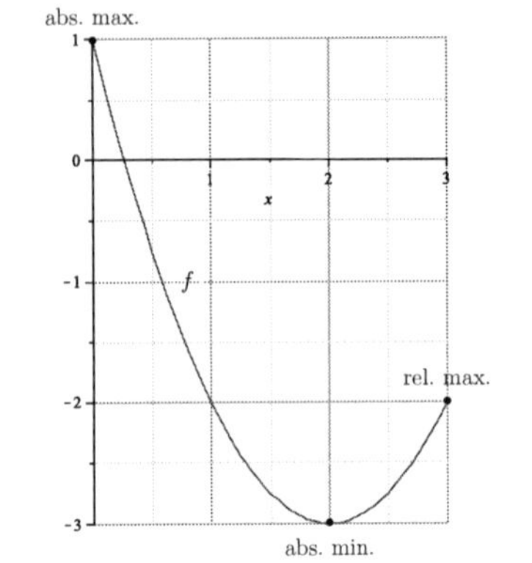

Figure 3.6:

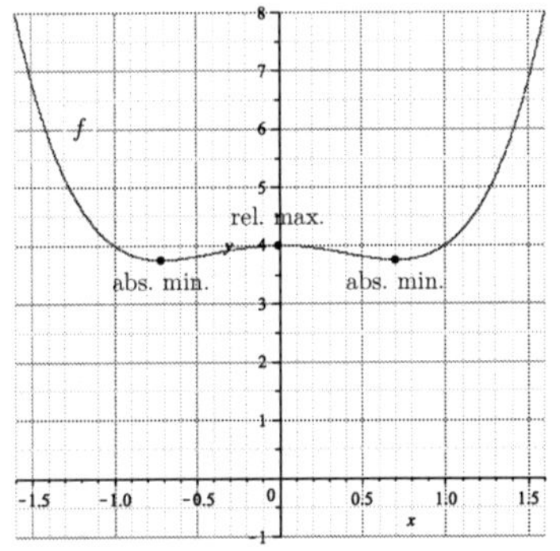

Figure 3.7:

3.4.13 Implicit Differentiation

1. Find dy/dx in terms of x and y:

   (a) $4\sqrt{x+y}\cdot\cos^2(\sqrt{x+y}) - 1$

   (b) ± 1

   (c) $-\dfrac{\sqrt[3]{y}}{\sqrt[3]{x}}$

   (d) $\dfrac{\cos^2(xy) - y}{x}$

   (e) $\dfrac{34x - 8xy + 3y}{8yx^2 - 3x}$

2. Find Equations for the Tangent and Normal Lines at the indicated Point, respectively:

   (a) $T(x) = -\frac{3}{2}x + 6;\quad N(x) = \frac{2}{3}x + \frac{5}{3}$

   (b) $T(x) = -x + 3; N(x) = x - 1$

   (c) $T(x) = -\frac{2}{3}\sqrt{3}\,x + \frac{\pi}{3} + \frac{\sqrt{3}}{3}$
   $N(x) = \frac{1}{2}\sqrt{3}\,x - \frac{1}{4}\sqrt{3} + \frac{\pi}{3}$

3.4.14 Extreme Values

Find the critical numbers and classify them as relative and absolute:

1. Critical numbers: 0, 2, 3; f has an absolute maximum at $(0,1)$, and absolute minimum at $(2,-3)$ and a relative maximum at $(3,-2)$.

2. Critical numbers: $\pm\frac{1}{2}\sqrt{2}$, 0; f has a relative maximum at $(0,4)$ and absolute minima at $\left(\pm\frac{1}{2}\sqrt{2}, \frac{15}{4}\right)$.

3. Critical numbers: 0, $\frac{12}{5}$, 4; f has an inflection point at $(0,0)$, a relative maximum at $(\frac{12}{5}, \frac{110\,592}{3\,125})$ and a relative minimum at $(4,0)$. Note that f does not have any absolute extrema, compare the Extreme Value Theorem!

4. Critical numbers: -2, 0, 6; f has a relative maximum at $(-2, \frac{4}{5})$, an absolute minimum at $(0,0)$ and an absolute maximum at $(6, \frac{36}{37})$.

5. Critical numbers: $\frac{1}{6}$; f has an absolute minimum at $\left(\frac{1}{6}, 24\right)$, but no absolute maximum. Note hat f does not satisfy the assumptions of the Extreme Value Theorem!

6. Critical numbers: -3, -2, 1; f has a relative maximum at $\left(-3, -\frac{3}{13}\right)$, an absolute minimum at $\left(-2, -\frac{1}{4}\right)$ and an absolute maximum at $\left(1, \frac{1}{5}\right)$.

7. critical numbers: -2, $-\sqrt{2}$, $\sqrt{2}$, 2; f has a relative maximum at $(-2,0)$, an absolute minimum at $(-\sqrt{2},-2)$, an absolute maximum at $(\sqrt{2},2)$ and a relative minimum at $(2,0)$.

8. critical points: 0, $\pi/4$, $\pi/2$, π; f has a relative minimum at $(0,2)$, its absolute maximum at $(\pi/2,3)$, its absolute minimum at $(\pi,-2)$ and an inflection point at $(\pi/4, 2\sqrt{2})$.

9. Critical numbers: -3, -1, 0, 2, 3, note that the domain of f is the half-open interval $[0,4)$; f has a relative maximum at $(-3,2)$, a relative minimum at $(-1,0)$, a relative maximum at $(0,2)$ and an absolute minimum at $(2,-2)$, however no absolute maximum. Note that f does not satisfy the assumption of the Extreme Value Theorem!

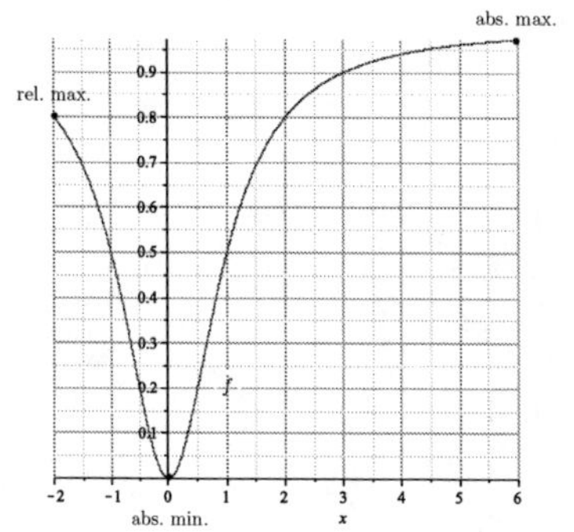

Figure 3.9:

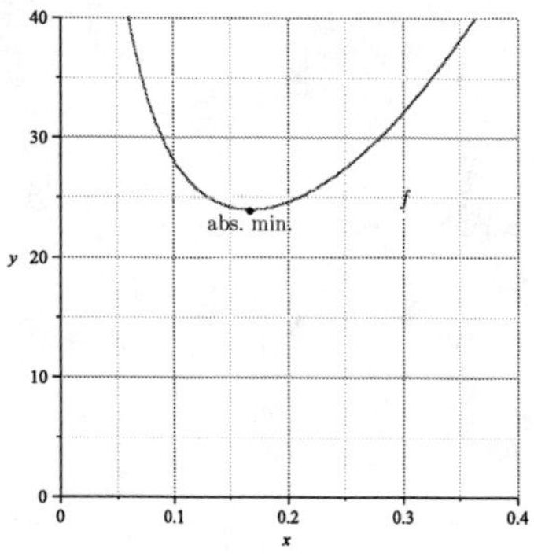

Figure 3.10:

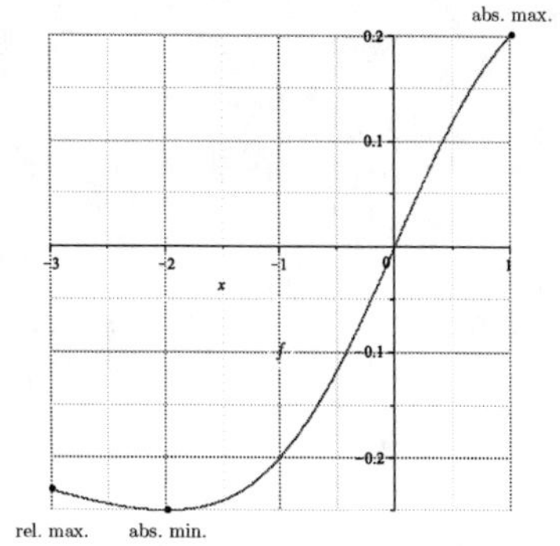

Figure 3.11:

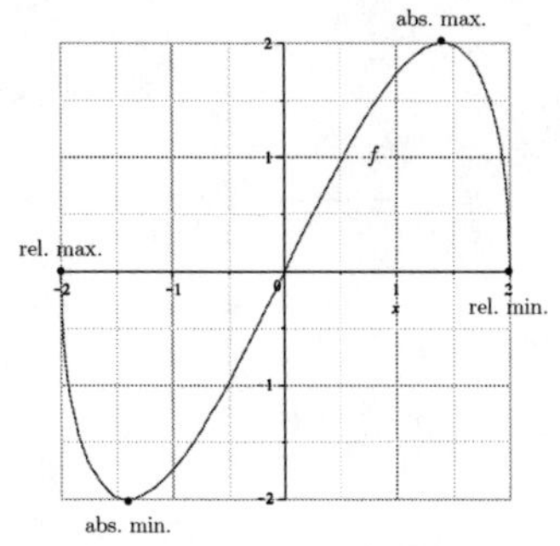

Figure 3.12:

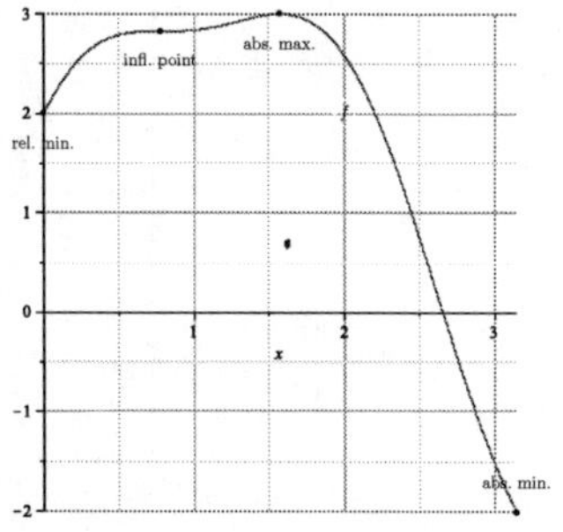

Figure 3.13:

3.4.15 Concavity, Inflection Points

Decribe the concavity of the graph of f and find the points of inflections:

1. f is concave down on $(-\infty, 0)$, concave up on $(0, \infty)$ and has an inflection point at $(0, 2)$.

2. f is concave down on $\left(-\infty, -\sqrt{3}\right) \cup \left(0, \sqrt{3}\right)$, concave up on $\left(-\sqrt{3}, 0\right) \cup \left(\sqrt{3}, \infty\right)$ and has an inflection points at $\left(-\sqrt{3}, -\frac{3}{2}\sqrt{3}\right)$, $(0, 0)$ and $\left(\sqrt{3}, \frac{3}{2}\sqrt{3}\right)$.

3. f is concave down on $\left(-\infty, -\frac{1}{2}\right)$, concave up on $\left(-\frac{1}{2}, \infty\right)$ and has an inflection point at $\left(-\frac{1}{2}, \frac{1}{2}\right)$.

4. f is concave up on $(-2, 0)$, concave down on $(0, 2)$, and has an inflection point at $(0, 0)$.

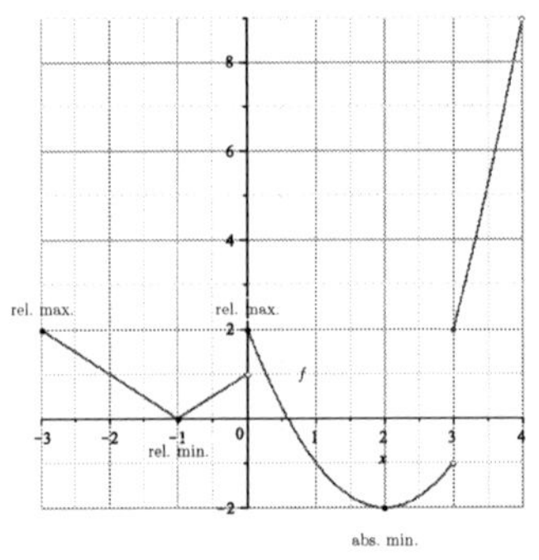

Figure 3.14:

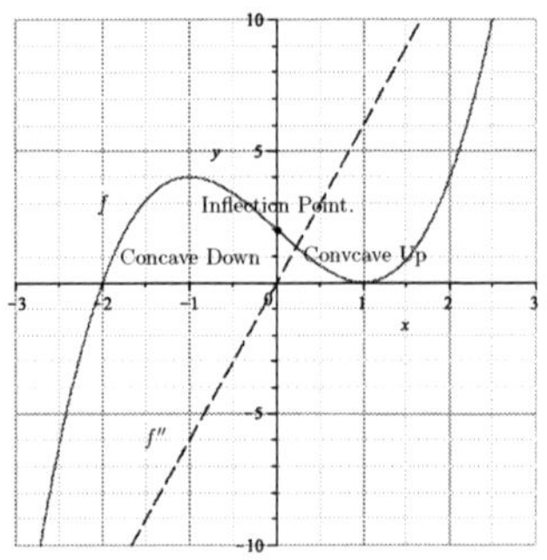

Figure 3.15:

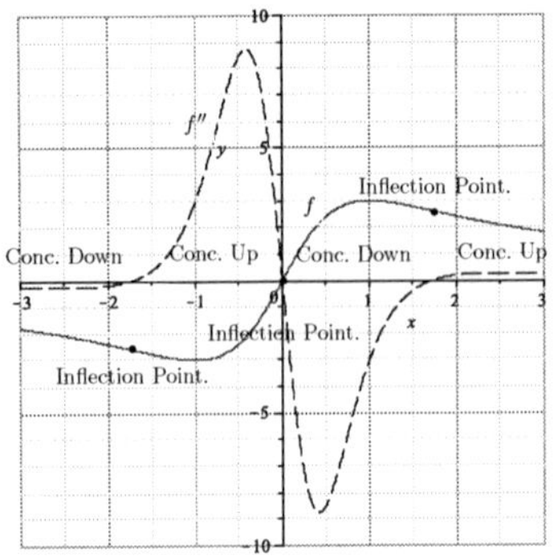

Figure 3.16:

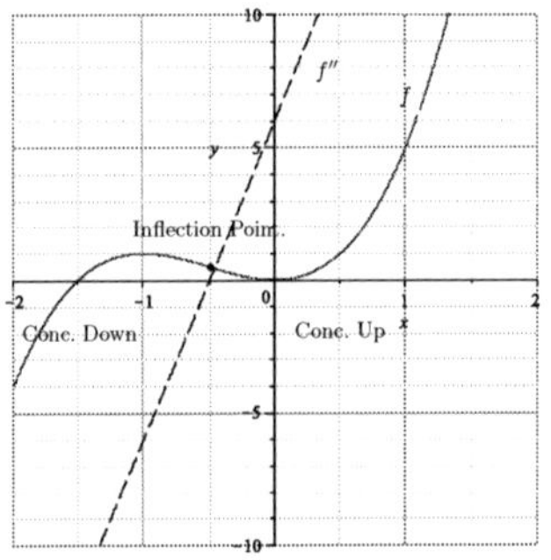

Figure 3.17:

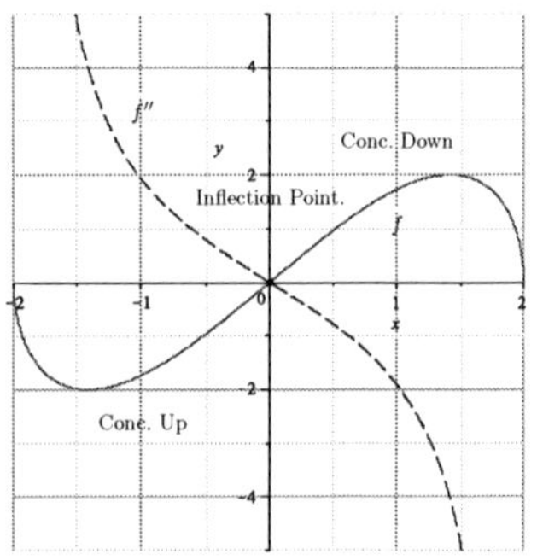

Figure 3.18:

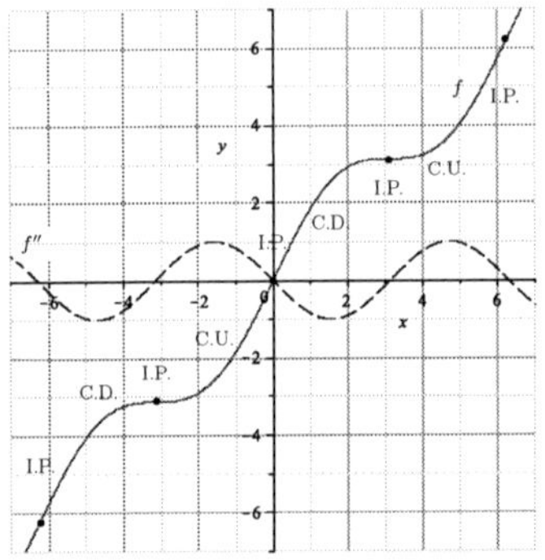

Figure 3.19:

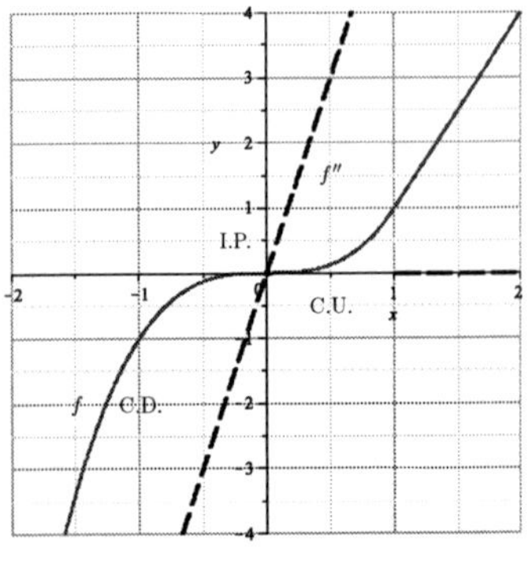

Figure 3.20:

5. f is concave up on the set

$$\bigcup_{n\in\mathbb{Z}} \Big((2n-1)\pi, 2n\pi\Big)$$

concave down on

$$\bigcup_{n\in\mathbb{Z}} \Big(2n\pi, (2n+1)\pi\Big)$$

and has inflection points at $(n\pi, n\pi)$ for $n \in \mathbb{Z}$.

6. f is concave down on $(-\infty, 0)$, concave up on $(0, 1)$, linear on $(1, \infty)$ and has an inflection point at $(0, 0)$

3.4.16 Monotonicity

Find the intervals on which f increases/decreases:

1. f is increasing on $(-\infty, -1) \cup (1, \infty)$, decreasing on $(-1, 0) \cup (0, 1)$, has a relative maximum at $(-1, -2)$ and a relative minimum at $(1, 2)$.

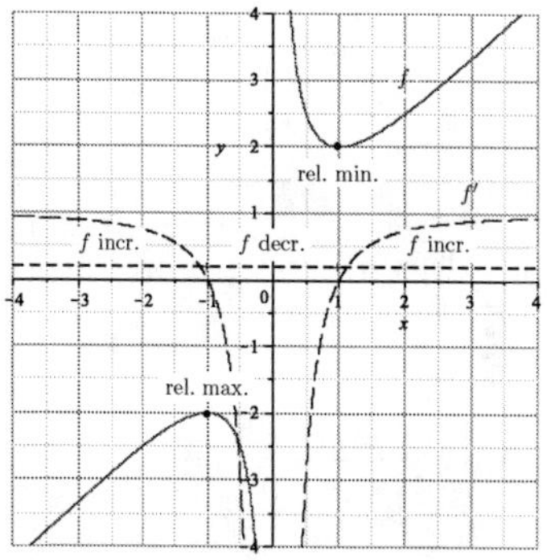

Figure 3.21:

2. f is increasing on $\big(-\infty, -\frac{14}{9}\big) \cup (0, \infty)$, decreasing on $(-\frac{14}{9}, 0)$, has a relative maximum at $\big(-\frac{14}{9}, \frac{3073}{243}\big)$ and a relative minimum at $(0, 7)$.

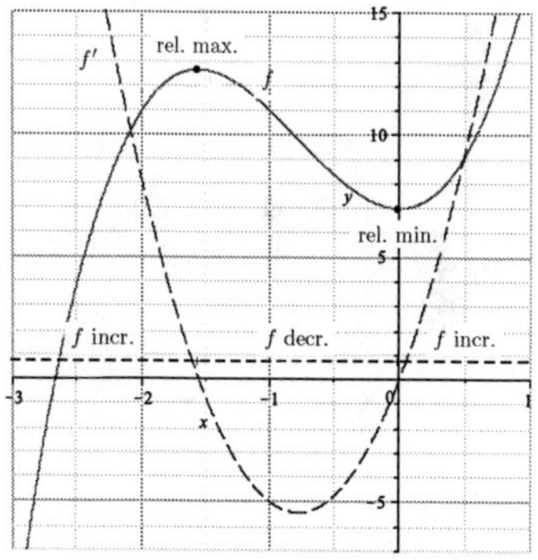

Figure 3.22:

3. Note hat f is only defined on $D_f := \mathbb{R} \setminus \{\pm 1\}$. f is increasing on $(-\infty, -1) \cup (-1, 0)$, decreasing on $(0, 1) \cup (1, \infty)$ and has a relative maximum at $(0, -1)$.

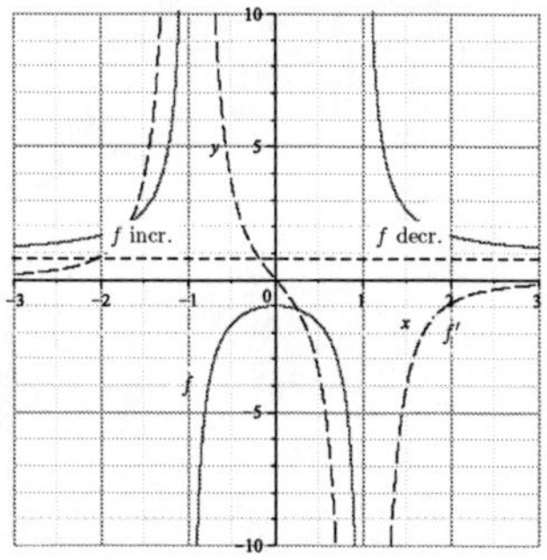

Figure 3.23:

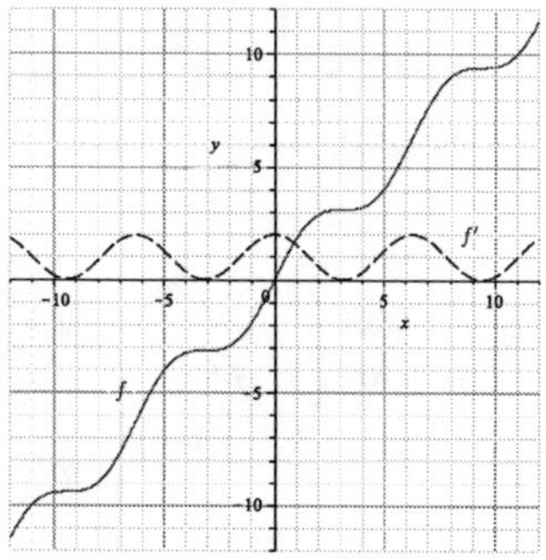

Figure 3.24:

4. f is increasing on $\mathbb{R}$. Note that f has inflection points at $\big(2(n-1)\pi, 2n-1)\pi\big)$ with horizontal tangent for $n \in \mathbb{Z}$.

3.4.17 Graph Sketching

Sketch the graph of the function f:

1. (i) Domain: $D_f := \mathbb{R}$, (ii) Symmetry: none, (iii) x-Intercepts: no rational zeroe, only one real zeroe $x \approx -0.4656$, (iv) y-Intercept:

$y = 1$, (v) Monotonicity: f is increasing on $\left(-\infty, \frac{1}{3}\right) \cup (1, \infty)$ and decreasing on $\left(\frac{1}{3}, 1\right)$, (vi) Extrema: f has a relative maximum at $\left(\frac{1}{3}, \frac{31}{27}\right)$ and a relative minimum at $(1, 1)$, but no absolute extremum, (vii) Concavity: f is concave down on $\left(-\infty, \frac{2}{3}\right)$ and concave up on $\left(\frac{2}{3}, \infty\right)$, (viii) Inflection Points: $\left(\frac{2}{3}, \frac{29}{27}\right)$, (ix) Asymptotes: none, $\lim_{x\to-\infty} f(x) = -\infty$, $\lim_{x\to\infty} f(x) = \infty$.

2. (i) Domain: $D_f := \mathbb{R} \setminus \{0\}$, (ii) Symmetry: none, (iii) x-Intercepts: $x = 4$, (iv) y-Intercept: none, (v) Monotonicity: f is decreasing on $(-\infty, 0) \cup (8, \infty)$ and decreasing on $(0, 8)$, (vi) Extrema: f has an absolute maximum at $\left(8, \frac{1}{16}\right)$, (vii) Concavity: f is concave down on $(-\infty, 0) \cup (0, 12)$ and concave up on $(12, \infty)$, (viii) Inflection Point: $\left(12, \frac{1}{18}\right)$, (ix) Asymptotes: $y = 0$ is a horizontal asymptote, $x = 0$ is a vertical asymptote of f.

3. (i) Domain: $D_f := \mathbb{R}\setminus\{-1\}$, (ii) Symmetry: none, (iii) x-Intercepts: $x = 0$, (iv) y-Intercepts: $y = 0$, (v) Monotonicity: f is increasing on $(-\infty, -2) \cup (0, \infty)$ and decreasing on $(-2, -1) \cup (-1, 0)$, (vi) Extrema: f has a relative maximum at $(-2, -8)$ and a relative minimum at $(0, 0)$, (vii) Concavity: f is concave down on $(-\infty, -1)$ and concave up on $(-1, \infty)$, (viii) Inflection Points: none, (ix) Asymptotes: Since

$$f(x) = \frac{2x^2}{x+1} = 2x - 2 + \frac{2}{x+1}$$

f has the (slanted) asymptote

$$y = 2x - 2.$$

In addition, f has the vertical asymptote $x = -1$.

4. (i) Domain: $D_f := \mathbb{R}$, (ii) Symmetry: even (iii) x-Intercepts: $x = \pm 2$, (iv) y-Intercept: $y = 2$, (v) Monotonicity: f is increasing on $(-\infty, -1) \cup (0, 1)$ and decreasing on $(-1, 0) \cup (1, \infty)$, (vi) Extrema: f has absolute maxima at $(-1, 3)$ and $(1, 3)$ and a relative minimum at $(0, 2)$, (vii) Concavity: f is concave down on $(-\infty, -1) \cup (1, \infty)$ and concave up on $(-1, 1)$, (viii) Inflection Points: $(-1, 3)$, $(1, 3)$, (ix) Asymptotes: none.

5. (i) Domain: $D_f := \mathbb{R}$, (ii) Symmetry: none, (iii) x-Intercepts: $x = 0, 1$, (iv) y-Intercept: $y = 0$, (v) Monotonicity: f is decreasing on $\left(-\infty, \frac{5}{6}\right)$ and increasing on $\left(\frac{5}{6}, \infty\right)$, (vi) Extrema: f has an absolute minimum at $\left(\frac{5}{6}, -\frac{5}{36} 6^{4/5}\right)$, (vii) Concavity: f is concave up on $(-\infty, 1) \cup \left(\frac{5}{3}, \infty\right)$ and concave down on $\left(1, \frac{5}{3}\right)$, (viii) Inflection Points: $(1, 0)$, and $\left(\frac{5}{3}, \frac{5}{9} \cdot 2^{1/5} \cdot 3^{4/5}\right)$, (ix) Asymptotes: none.

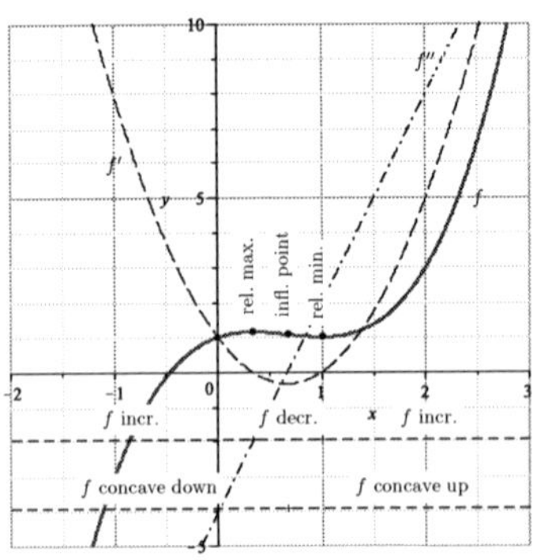

Figure 3.25:

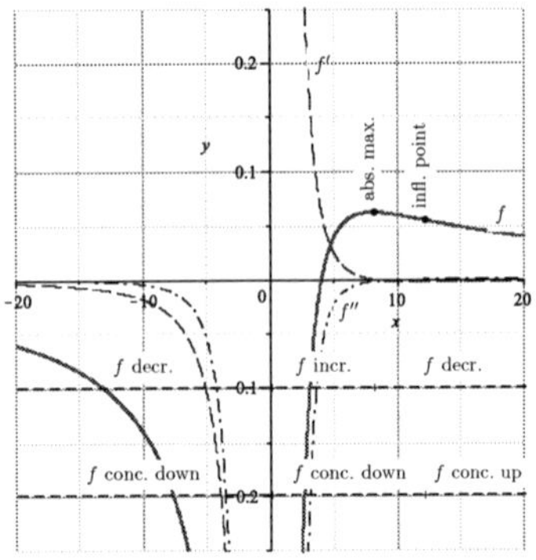

Figure 3.26:

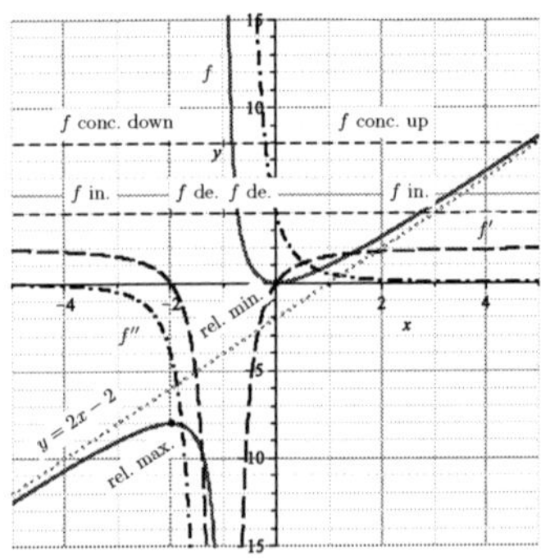

Figure 3.27:

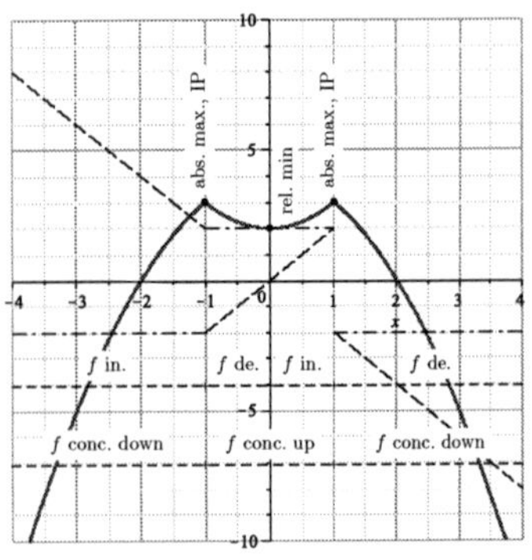

Figure 3.28:

3.4.18 Maximum/Minimum Problems

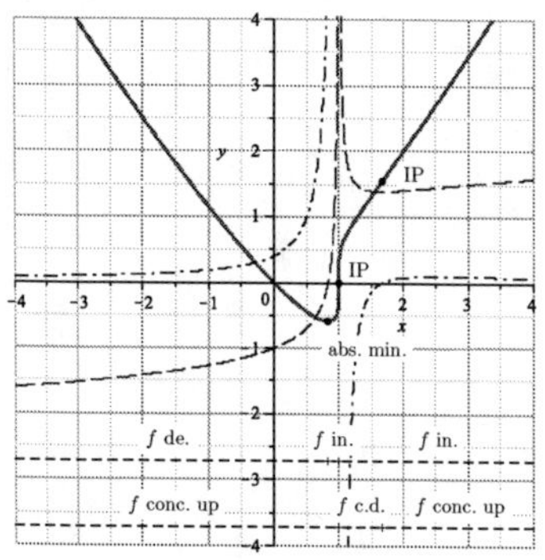

Figure 3.29:

1. The triangle has to be equilateral, with each side equaling 4 units!
2. $\pm\left(\sqrt{\frac{5}{2}}, \frac{5}{2}\right)$
3. $4\sqrt{6}$ ft.
4. Width: 265 ft (occurs 2 times), depth: 106 ft (occurs 5 times!).
5. $\frac{3}{4}\sqrt{3}\,r^2$ units^2
6. Suppose the conical cup has radius r and height h and volume V. We would like to minimize its (surface) area A and will show that A attains its absolute minimum at $h = \sqrt{2}r$.

   Recall that
   $$V = \tfrac{1}{3}\pi r^2 h$$
   and thus
   $$h = \frac{3V}{\pi r^2}.$$
   Moreover,
   $$A = \pi r\sqrt{r^2 + h^2} = \pi\left(r^2 + \frac{9V^2}{\pi^2 r^4}\right)$$
   $$= \left(\frac{\pi^2 r^6 + 9V^2}{r^2}\right).$$
   Note that A gives the surface area of the conical cup of radius $r > 0$ with fixed volume V. In order to find its minimum, we first compute its derivative
   $$A'(r) = \frac{\dfrac{6\pi^2 r^7 - 2r(\pi^2 r^6 + 9V^2)}{r^4}}{2\sqrt{\dfrac{\pi^2 r^6 + 9V^2}{r^2}}}$$
   $$= \frac{4\pi^2 r^7 - 18rV^2}{2r^4\sqrt{\dfrac{\pi^2 r^6 + 9V^2}{r^2}}} = \frac{2\pi^2 r^6 - 9V^2}{r^3\sqrt{\dfrac{\pi^2 r^6 + 9V^2}{r^2}}}$$
   It can be easily seen that $A'(r) = 0$ if and only if $r^3 = \frac{3V}{\sqrt{2}\pi}$ and that A has an absolute minimum at this point. Hence, since $3V = h\pi r^2$,
   $$r^3 = \frac{3V}{\sqrt{2}\pi} = \frac{h\pi r^2}{\sqrt{2}, \pi} = \frac{hr^2}{\sqrt{2}}$$
   which is equivalent to $r\sqrt{2} = h$, as claimed.
7. 4 ft^2

3.5 Solutions to Selected Exercises

3.5.1 Two-Sided Limits

Find the following limits if they exist:

1. We first reduce the argument of the limit operator and obtain the expression

$$\lim_{x\to c} \frac{2x-6}{x-3} = \lim_{x\to c} \frac{2(x-3)}{x-3} = \lim_{x\to c} 2$$

   whose limit clearly equals

$$= 2$$

   by Theorem 3.2:1.

2. Let $q(x) := x^2 - 6x - 9$. As can be easily verified, $q(3) = -18 \neq 0$, hence $3 \in D_r$, where $r(x) = (x-3)/(x^2-6x-9)$. Hence, we can evaluate the limit of $r(x)$ at the point $c = 3$ using Theorem 3.2:4 and obtain

$$\lim_{x\to 3} \frac{x-3}{x^2-6x-9} = \frac{3-3}{3^2 - 6\cdot 3 - 9} = 0.$$

3. Since the polynomial $q(x) = x^2 - 5x + 4$ is zero at $c = 1$, as can be easily verified, we cannot apply Theorem 3.2:4 to compute the given limit. Therefore, we try to reduce the fraction:

$$\lim_{x\to 1} \frac{x-1}{x^2-5x+4} = \lim_{x\to 1} \frac{x-1}{(x-1)(x-4)} = \lim_{x\to 1} \frac{1}{x-4}$$

   The denominator of the reduced fraction is different from zero at $c = 1$. Hence, we can apply Theorem 3.2:4 and obtain

$$= \frac{1}{1-4} = -\tfrac{1}{3}$$

4. Expanding the fraction by x yields

$$\lim_{x\to 0} \frac{1-\frac{1}{x}}{1+\frac{1}{x}} = \lim_{x\to 0} \frac{x\left(1-\frac{1}{x}\right)}{x\left(1+\frac{1}{x}\right)} = \lim_{x\to 0} \frac{x-1}{x+1}$$

   which satisfies the assumptions of Theorem 3.2:4. Hence the limit can be determined by evaluating the argument function at $c = 0$

$$= \frac{0-1}{0+1} = -1$$

5. Factoring the numerator and reducing the quotient, we obtain

$$\begin{aligned}\lim_{x\to1}\frac{3x-3}{\sqrt{x-1}} &= \lim_{x\to1}\frac{3(x-1)}{(x-1)^{1/2}}\\ &= 3\lim_{x\to1}(x-1)^{1/2} = 3\cdot0 = 0\end{aligned}$$

by the Substitution Rule for limits and the continuity of the root functions.

6. We combine the fractions in the denominator and factor the denominator using the identity $x-2=\left(\sqrt{x}-\sqrt{2}\right)\left(\sqrt{x}+\sqrt{2}\right)$, then follows

$$\begin{aligned}\lim_{x\to2}\frac{\frac{1}{\sqrt{2}}-\frac{1}{\sqrt{x}}}{x-2} &= \lim_{x\to2}\frac{\frac{\sqrt{x}-\sqrt{2}}{\sqrt{2x}}}{\left(\sqrt{x}-\sqrt{2}\right)\left(\sqrt{x}+\sqrt{2}\right)}\\ &= \lim_{x\to2}\frac{\sqrt{x}-\sqrt{2}}{\sqrt{2x}\left(\sqrt{x}-\sqrt{2}\right)\left(\sqrt{x}+\sqrt{2}\right)}\\ &= \lim_{x\to2}\frac{1}{\sqrt{2x}\left(\sqrt{x}+\sqrt{2}\right)}\end{aligned}$$

The limit of the denominator can be computed using the product rule and summation rule for limits (Theorem 3.4:3 and 1) in conjunction with Theorem 3.3:1 and Theorem 3.2:3

$$\begin{aligned}\lim_{x\to2}\sqrt{2x}\left(\sqrt{x}+\sqrt{2}\right) &= \lim_{x\to2}\sqrt{2x}\left(\lim_{x\to2}\sqrt{x}+\lim_{x\to2}\sqrt{2}\right)\\ &= \sqrt{\lim_{x\to2}2x}\left(\sqrt{\lim_{x\to2}x}+\sqrt{2}\right)\\ &= \sqrt{2\cdot2}\left(\sqrt{2}+\sqrt{2}\right) = 4\sqrt{2}\neq0.\end{aligned}$$

Hence, by the quotient rule (Theorem 3.4:4)

$$\lim_{x\to2}\frac{1}{\sqrt{2x}\left(\sqrt{x}+\sqrt{2}\right)} = \frac{\lim_{x\to2}1}{\lim_{x\to2}\sqrt{2x}\left(\sqrt{x}+\sqrt{2}\right)} = \frac{1}{4\sqrt{2}} = \tfrac{1}{8}\sqrt{2}.$$

Hence,

$$\lim_{x\to2}\frac{\frac{1}{\sqrt{2}}-\frac{1}{\sqrt{x}}}{x-2} = \tfrac{1}{8}\sqrt{2}.$$

7. We first factor and reduce the argument of the limit operator:

$$\begin{aligned}\lim_{x\to1}\frac{x^{2/3}-1}{x-1} &= \lim_{x\to1}\frac{\left(x^{1/3}-1\right)\left(x^{1/3}+1\right)}{\left(x^{1/3}-1\right)\left(x^{2/3}+x^{1/3}+1\right)}\\ &= \lim_{x\to1}\frac{x^{1/3}+1}{x^{2/3}+x^{1/3}+1}\end{aligned}$$

Since by Theorem 3.4:3, Theorem 3.3:1 and Theorem 3.2:3

$$\lim_{x\to 1}\left(x^{2/3}+x^{1/3}+1\right) = \left(\lim_{x\to 1} x\right)^{2/3} + \left(\lim_{x\to 1} x\right)^{1/3} + \lim_{x\to 1} 1$$
$$= 1+1+1 = 3 \neq 0,$$

we can use the quotient rule (Theorem 3.4:4) to finish evaluating the limit above. Therefore, by another application of Theorem 3.3:1 and Theorem 3.2:3, we obtain

$$\lim_{x\to 1}\frac{x^{1/3}+1}{x^{2/3}+x^{1/3}+1} = \frac{\lim_{x\to 1}\left(x^{1/3}+1\right)}{\lim_{x\to 1}\left(x^{2/3}+x^{1/3}+1\right)}$$
$$= \frac{\left(\lim_{x\to 1} x\right)^{1/3}+1}{3} = \frac{2}{3}.$$

Hence,

$$\lim_{x\to 1}\frac{x^{2/3}-1}{x-1} = \tfrac{2}{3}.$$

8. Clearly,

$$\lim_{x\to 0}\frac{\tan^2 x}{x} = \lim_{x\to 0}\left(\frac{\sin x}{x}\cdot\frac{\sin x}{\cos^2 x}\right)$$

and by the product rule (Theorem 3.4:3)

$$= \lim_{x\to 0}\frac{\sin x}{x}\cdot\lim_{x\to 0}\frac{\sin x}{\cos^2 x}$$

Note that both limits exist by Theorem 3.3:6,2,3. Moreover, by Theorem 3.4:4 and 3, we have

$$\lim_{x\to 0}\frac{\sin x}{\cos^2 x} = \frac{\lim_{x\to 0}\sin x}{\left(\lim_{x\to 0}\cos x\right)^2} = \frac{0}{1} = 0.$$

Therefore,

$$\lim_{x\to 0}\frac{\sin x}{x}\cdot\lim_{x\to 0}\frac{\sin x}{\cos^2 x} = \lim_{x\to 0}\frac{\sin x}{x}\cdot\lim_{x\to 0}\frac{\sin x}{\cos^2 x} = 1\cdot 0 = 0,$$

hence

$$\lim_{x\to 0}\frac{\tan^2 x}{x} = 0.$$

9. Rewriting the limit using the substitution $y = x - \pi$, we see that

$$\lim_{x\to\pi} \frac{\sin x}{x-\pi} = \lim_{y\to 0} \frac{\sin(y+\pi)}{y}$$

which, by the addition formula for sin, equals

$$= \lim_{y\to 0} \frac{\sin y \cos\pi + \sin\pi \cos y}{y}$$

and, since $\sin\pi = 0$, $\cos\pi = -1$,

$$= \lim_{y\to 0} \frac{-\sin y}{y}$$

and

$$= -\lim_{y\to 0} \frac{\sin y}{y} = -1$$

by Theorem 3.4:2 and Theorem 3.3:6.

10. Clearly,

$$\begin{aligned}\lim_{x\to 0} \frac{\sin 2x}{\sin 3x} &= \lim_{x\to 0} \frac{(\sin 2x)2x}{2x\sin 3x} \\ &= \lim_{x\to 0} \frac{(\sin 2x)3x\frac{2}{3}}{2x\sin 3x} = \lim_{x\to 0}\left(\frac{\frac{\sin 2x}{2x}}{\frac{\sin 3x}{3x}}\cdot\frac{2}{3}\right)\end{aligned}$$

which by Theorem 3.4:4,2 equals

$$= \frac{2}{3}\frac{\lim_{x\to 0}\frac{\sin 2x}{2x}}{\lim_{x\to 0}\frac{\sin 3x}{3x}}$$

and by Theorem 3.3:6

$$= \tfrac{2}{3}\cdot\frac{1}{1} = \tfrac{2}{3}$$

11. We compute this limit using L'Hôpital's Rule (Theorem 3.8). Note that numerator and denominator function are differentiable at $\frac{\pi}{4}$ and the limit is of type $\frac{0}{0}$:

$$\lim_{x\to\pi/4} \frac{\tan(\pi-4x)}{x-\pi/4} = \lim_{x\to\pi/4} \frac{-4\sec^2(\pi-4x)}{1} = -4.$$

12. We will use the "Squeeze Theorem" (Theorem 3.7): Since

$$-1 \leq \cos x \leq 1, \quad \text{for all } x \in \mathbb{R}$$

it follows by multiplying the double-inequality by $|x| \geq 0$ that

$$-|x| \leq |x| \cos x \leq |x| \quad \text{for all } x \in \mathbb{R}.$$

Since by Theorem 3.3:1 and Theorem 3.4:3

$$\lim_{x\to 0} |x| = \lim_{x\to 0} \sqrt{x^2} \sqrt{\left(\lim_{x\to 0} x\right)^2} = \sqrt{0} = 0,$$

by Theorem 3.7, the limit of the expression in the middle of the double-inequality above has to be zero,too, i.e.

$$\lim_{x\to 0} |x| \cos x = 0.$$

3.5.2 One-Sided Limits

Find the following limits if they exist:

2. Clearly,

$$\frac{\sin x}{\sqrt{x}} = \frac{\sin x}{x} \cdot \sqrt{x}$$

for all $x \in \mathbb{R}$, $x \geq 0$. Hence by Theorem 3.4:3

$$\lim_{x\to 0^+} \frac{\sin x}{\sqrt{x}} = \lim_{x\to 0^+} \left(\frac{\sin x}{x} \cdot \sqrt{x}\right) = \lim_{x\to 0^+} \frac{\sin x}{x} \lim_{x\to 0^+} \sqrt{x}$$

which by Theorem 3.3:6 and 1 equals

$$\begin{aligned} &= 1 \cdot \sqrt{\lim_{x\to 0^+} x} \\ &= 1 \cdot 0 = 0 \end{aligned}$$

by Theorem 3.2:2.

3.5.3 Limits at Infinity

Find the following limits if they exist:

2. Clearly,

$$\lim_{x\to\infty} \frac{x+1}{x-3} = \lim_{x\to\infty} \frac{x^2\left(1+\frac{1}{x^2}\right)}{x^2\left(2+\frac{4}{x^2}\right)} = \lim_{x\to\infty} \frac{1+\frac{1}{x^2}}{2+\frac{4}{x^2}}$$

which in turn by Theorem 3.4:4, 1 and 3 equals

$$= \frac{\lim\limits_{x\to\infty} 1 + \left(\lim\limits_{x\to\infty} \frac{1}{x}\right)^2}{\lim\limits_{x\to\infty} 2 + \left(\lim\limits_{x\to\infty} \frac{1}{x}\right)^2}$$

and by Theorem 3.2:1 and Theorem 3.3:8 finally

$$= \frac{1+0^2}{2+0^2} = \tfrac{1}{2}.$$

3.5.4 Squeeze Play

Find the following limits if they exist.

3. Clearly, $e/n \leq 1$ for all $n \in \mathbb{N}$ with $n \geq 3$. Hence

$$\begin{aligned}\frac{e^n}{n!} &= \frac{1}{e}\frac{e}{2}\frac{e}{3}\frac{e}{4}\cdots\frac{e}{n-1}\frac{e}{n} = \frac{e^3}{2}\frac{e}{3}\frac{e}{4}\cdots\frac{e}{n-1}\frac{1}{n}\\ &\leq \frac{e^3}{2}\cdot 1\cdot 1\cdots 1\cdot\frac{1}{n} = \frac{e^3}{2}\frac{1}{n}\end{aligned}$$

for all $n \in \mathbb{N}$ with $n \geq 4$. Therefore, we have established that

$$0 \leq \frac{e^n}{n!} \leq \frac{e^3}{2}\frac{1}{n}$$

for all $n \in \mathbb{N}$ with $n \geq 4$. Since

$$\lim_{n\to\infty}\frac{e^3}{2}\frac{1}{n} = \frac{e^3}{2}\lim_{n\to\infty}\frac{1}{n} = \frac{e^3}{2}\cdot 0 = 0$$

by Theorem 3.4:2 and Theorem 3.3:8, we can apply the "Squeezing Theorem" (Theorem 3.7) and obtain that the middle expression in the double-inequality above also has to be zero, i.e.

$$\lim_{n\to\infty}\frac{e^n}{n!} = 0.$$

3.5.5 L'Hôpital's Rule

Find the following limits if they exist:

1. Since $f(x) := 3^x$ and $g(x) := \ln x$ are differentiable in a neigborhood of $c = \infty$ and $\lim_{x\to]infty} 3^x = \lim_{x\to\infty} e^{x\ln x} = \infty$ by Theorem 3.3:10 and $\lim_{x\to\infty}\ln x = \infty$ by Theorem 3.3:11, we can apply L'Hôpital's Rule (Theorem 3.8) and conclude

$$\lim_{x\to\infty}\frac{3^x}{\ln x} = \lim_{x\to\infty}\frac{3^x\ln 3}{\frac{1}{x}} = \lim_{x\to\infty}\left(x3^x\ln 3\right) = \infty.$$

2. We first rewrite the argument of the limit operator in terms of the natural exponential function:

$$\left(1+\frac{a}{x}\right)^x = e^{x\ln(1+\frac{a}{x})}.$$

We now apply the Substitution Rule (Theorem 3.5) to this expression and first compute the limit of the interior function $f(x) := x\ln(1+\frac{a}{x})$:

$$\lim_{x\to\infty} x\ln(1+\frac{a}{x}) = \lim_{x\to\infty}\frac{\ln(1+\frac{a}{x})}{\frac{1}{x}}$$

which satisfies the assumptions of L'Hôpital's Rule (Theorem 3.8) as can be easily verified and therefore equals

$$\begin{aligned} &= \lim_{x\to\infty}\left(\frac{\frac{x}{x+a}(-\frac{a}{x^2})}{-\frac{1}{x^2}}\right) = \lim_{x\to\infty}\frac{ax}{x+a} \\ &= \lim_{x\to\infty}\frac{x\cdot a}{x(1+\frac{a}{x})} = \lim_{x\to\infty}\frac{a}{1+\frac{a}{x}} = \frac{a}{1+0} = a \end{aligned}$$

by Theorem 3.4:4, Theorem 3.3:8. By Theorem 3.3:4,

$$\lim_{y\to a} e^y = e^a,$$

hence by the Substitution Rule (Theorem 3.5)

$$\lim_{x\to\infty} e^{x\ln(1+\frac{a}{x})} = e^a.$$

Therefore,

$$\lim_{x\to\infty}\left(1+\frac{a}{x}\right)^x = e^a.$$

3.5.6 Continuity, Removable Dicontinuities, etc.

1. f is continuous on $\mathbb{R}\setminus\{0,1,2\}$. We will check the half-sided limits at 0, 1 and 2:

$$\begin{aligned} \lim_{x\to 0^-} f(x) &= \lim_{x\to 0^-} 1 = 1 \\ \lim_{x\to 0^+} f(x) &= \lim_{x\to 0^+} x^2 = 0, \end{aligned}$$

hence f has a jump discontinuity at $x=0$. Moreover

$$\begin{aligned} \lim_{x\to 1^-} f(x) &= \lim_{x\to 1^-} x^2 = 1 \\ \lim_{x\to 1^+} f(x) &= \lim_{x\to 1^+} 1 = 1, \end{aligned}$$

hence f has a removable discontinuity at $x = 1$. Finally

$$\lim_{x\to 2^-} f(x) = \lim_{x\to 2^-} 1 = 1$$
$$\lim_{x\to 2^+} f(x) = \lim_{x\to 2^+} x = 2,$$

hence f has a jump discontinuity at $x = 2$.

2. f is clearly continuous on $\mathbb{R}\setminus\{-3,-1,3\}$. We will check the half-sided limits at -3, -1 and 1:

$$\lim_{x\to -3^-} f(x) = \lim_{x\to -3^-} (x+7) = 4$$
$$\lim_{x\to -3^+} f(x) = \lim_{x\to -3^+} |x-2| = 5,$$

hence f has a jump discontinuity at $x = -3$. Moreover

$$\lim_{x\to -1^-} f(x) = \lim_{x\to -1^-} |x-2| = 3$$
$$\lim_{x\to -1^+} f(x) = \lim_{x\to -1^+} (x^2-2x) = 3$$

hence f has a removable discontinuity at $x = -1$. Note that f is not defined at $x = -1$. Finally

$$\lim_{x\to 3^-} f(x) = \lim_{x\to 3^-} (x^2-2x) = 3$$
$$\lim_{x\to 3^+} f(x) = \lim_{x\to 3^+} (2x-3) = 3 = f(3)$$

hence f is continuous at $x = 3$.

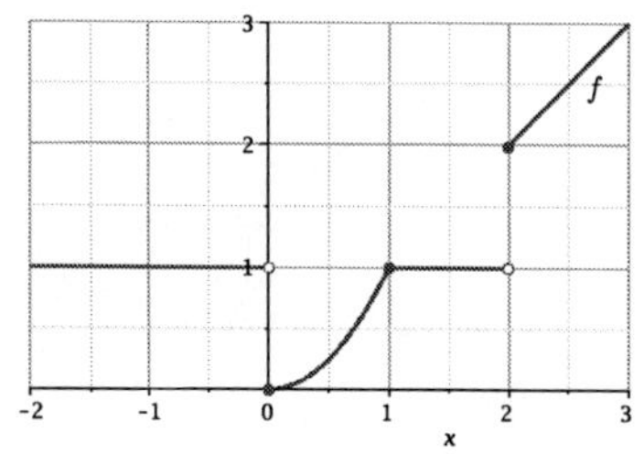

Figure 3.30:

3. The domain of f is $D_f := \mathbb{R}\setminus\{-1,-4\}$. Hence, als a rational function, f is continuous on D_f. However, since

$$f(x) = \frac{x+1}{x^2+5x+4} = \frac{x+1}{(x+1)(x+4)} = \frac{1}{x+4},$$

we see that f has a removable discontinuity at $x = -1$.

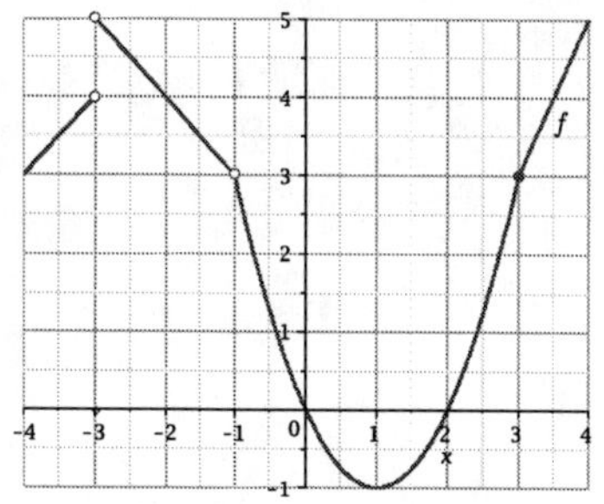

Figure 3.31:

4. As a rational function, f is continuous on its domain $D_f := \mathbb{R}\setminus\{3\}$. We check the half-sided limits at $x = 3$:

$$\lim_{x\to 3^-} f(x) = \lim_{x\to 3^-} \frac{x+3}{x-3} = -\infty$$

Note that $x \to 3^-$ implies that $x - 3 < 0$ and $x + 3 > 0$. Hence the quotient is negative as $x \to 3^-$ with the numerator being bounded and the denominator approaching zero. Similarly, we find that

$$\lim_{x\to 3^+} f(x) = \lim_{x\to 3^+} \frac{x+3}{x-3} = +\infty$$

Again, note that $x \to 3^+$ implies that $x - 3 > 0$ and $x + 3 > 0$ etc.

5. Since

$$f(x) = \frac{x-2}{x^2-4} = \frac{x-2}{(x-2)(x+2)} = \frac{1}{x+2},$$

f has a removable discontinuity at $x = 2$. Thus, if we define

$$f(x) := \lim_{x \to 2} \frac{1}{x+2} = \tfrac{1}{4},$$

(the extension of) f is continuous at $x = 2$.

3.5.7 Intermediate Value Theorem

Recall and interpret the Intermediate Value Theorem. The real valued, continuous function f is defined on the closed and bounded interval $[a, b]$. Which of the following must be true?

3.5.8 Mean Value Theorem

Select the correct formulation for the Mean Value Theorem:

3.5.9 Extreme Value Theorem

The real valued, continuous function f is defined on the closed and bounded interval $[a, b]$. Which of the following must be true?

3.5.10 Equations of Tangents and Normals

Find the equation for the tangent and normal to the given curve at the given point:

1. Clearly, the derivative of f is given by

$$f'(x) = \cos x$$

for $x \in \mathbb{R}$. The slope of f at $(\frac{\pi}{6}, \frac{1}{2})$, which by definition is the slope of the tangent at $(\frac{\pi}{6}, \frac{1}{2})$, is therefore $m = f'(\frac{\pi}{6}) = \cos\frac{\pi}{6} = \frac{1}{2}\sqrt{3}$. With the point-slope form of the line equation, which is given by

$$y - y_0 = m(x - x_0),$$

we can now easily compute he tangent line at the point $(\frac{\pi}{6}, \frac{1}{2})$:

$$y - \tfrac{1}{2} = \tfrac{1}{2}\sqrt{3}(x - \tfrac{\pi}{6}) = \tfrac{1}{2}\sqrt{3}x - \tfrac{\pi}{12}\sqrt{3}.$$

Hence the tangent line is given by

$$T(x) = (\tfrac{1}{2}\sqrt{3})x + (\tfrac{1}{2} - \tfrac{\pi}{12}\sqrt{3})$$

for all $x \in \mathbb{R}$.

The normal line of f at $(\frac{\pi}{6}, \frac{1}{2})$, can be computed the same way. Recall that the slope of the normal at $(\frac{\pi}{6}, \frac{1}{2})$ is given by

$$\begin{aligned} m^{\perp} &= -\frac{1}{m} = -\frac{1}{f'(\frac{\pi}{6})} \\ &= -\frac{1}{\frac{1}{2}\sqrt{3}} = -\frac{2}{\sqrt{3}} = -\tfrac{2}{3}\sqrt{3}. \end{aligned}$$

We can now compute the normal line of f at the point $(\frac{\pi}{6}, \frac{1}{2})$:

$$y - \tfrac{1}{2} = -\tfrac{2}{3}\sqrt{3}(x - \tfrac{\pi}{6}) = -\tfrac{2}{3}\sqrt{3}x + \tfrac{\pi}{9}\sqrt{3}$$

and obtain

$$N(x) = (-\tfrac{2}{3}\sqrt{3})x + (\tfrac{1}{2} + \tfrac{\pi}{9}\sqrt{3})$$

for all $x \in \mathbb{R}$.

2. Since the derivative of f is

$$f(x) = \frac{1}{2\sqrt{x+1}}$$

the slope of f at $(3, 2)$ is

$$m = f'(3) = \frac{1}{2\sqrt{3+1}} = \tfrac{1}{4}.$$

Moreover, the slope of the normal line at $(3, 2)$ is given by

$$m^{\perp} = -\frac{1}{m} = -\frac{1}{\frac{1}{4}} = -4.$$

With the point-slope form $y - y_0 = m(x - x_0)$ of the line equation, we obtain $y - 2 = \frac{1}{4}(x - 3) = \frac{1}{4}x - \frac{3}{4}$ and $y - 2 = -4(x - 3) = -4x + 12$. Thus, the tangent line of f at $(3, 2)$ is given by

$$T(x) = \tfrac{1}{4}x + \tfrac{5}{4}$$

while the normal line of f at $(3, 2)$ is

$$N(x) = -4x + 14$$

for $x \in \mathbb{R}$.

3. The derivative of f is $f'(x) = e^{-x}$, hence the slope of the tangent of f at the point $(0, 0)$ is
$$m = f'(0) = e^{-0} = 1$$
while the slope of the normal of f at $(0, 0)$ is
$$m^{\perp} = -\frac{1}{m} = -\frac{1}{1} = -1.$$
We can therefore compute the equations of the tangent and normal line with point-slope form of the line equation and obtain $y - 0 = 1 \cdot (x - 0) = x$ and $y - 0 = -1 \cdot (x - 0) = -x$, respectively. Hence, the tangent line and normal line of f at $(0, 0)$ are given by
$$T(x) = x$$
and
$$N(x) = -x,$$
respectively.

4. The derivative of f is $f'(x) = \frac{1}{x}$, hence the slope of the tangent and normal of f at the point $(3, 1 + \ln 3)$ is
$$m = f'(3) = \tfrac{1}{3}$$
and
$$m^{\perp} = -\frac{1}{m} = -\frac{1}{\frac{1}{3}} = -3,$$
respectively. With the point-slope form of the line equation, we again obtain $y - (1 + \ln 3) = \frac{1}{3}(x - 3) = \frac{1}{3}x - 1$ and $y - (1 + \ln 3) = -3(x - 3) = -3x + 9$. Hence
$$T(x) = \tfrac{1}{3}x + \ln 3$$
is the equation of the tangent line of f at the $(3, 1 + \ln 3)$, while the eation of the normal line of f at the same point is given by
$$N(x) = -3x + (10 + \ln 3)$$
for $x \in \mathbb{R}$¿

3.5.11 The Derivative

2. Use the Definition to compute the Derivative of the following Functions at x_0:

7. Determine whether each of the following function is (α) differentiable (β) continuous at x_0:

3.5.12 Applying the Differentation Rules

Rational Functions

Trigonometric Functions

Exponential and Logarithmic Functions

Miscellaneous

3.5.13 Implicit Differentiation

1. Find dy/dx in terms of x and y:

7. Find Equations for the Tangent and Normal Lines at the indicated Point, respectively:

3.5.14 Extreme Values

Find the critical numbers and classify them as relative and absolute:

1. (i) Find the critical numbers of f on $(0, 3)$: $f'(x) = 2x - 4 = 2(x - 2) = 0$ which is equivalent to $x = 2$; note $2 \in (0, 3)$. Hence, the critical numbers of f on $[0, 3]$ are $\{0, 2, 3\}$.

   (ii) Determine the type of the critical numbers: We will apply the First Derivative Test. Pick a sample point in each subinterval determined by the critical numbers. We choose 1 and $5/2$.

   Then evaluate f' at those sample points. Actually, we only need to determine the sign (postive/negative). We find that $f'(1) < 0$ and $f'(5/2) > 0$. By the Intermediate-Value Theorem, f' keeps its sign on the whole subinterval. Thus, by the First Derivative Test, f has a relative maximum at 0, a relative minimum at 2 and a relative maximum at 3.

   (iii) Evaluate f at each extremum and compare the values: $f(0) = 1$, $f(2) = 4 - 8 + 1 = -3$, $f(3) = 9 - 12 + 1 = -2$. Hence $(0, 1)$ is the absolute maximum, $(2, -3)$ the absolute minimum and $(3, -2)$ a relative maximum of f on $[0, 3]$.

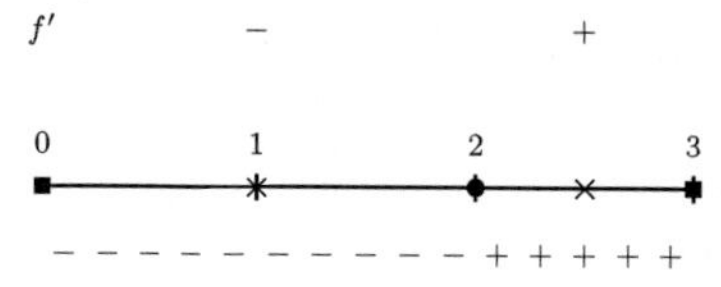

Figure 3.32:

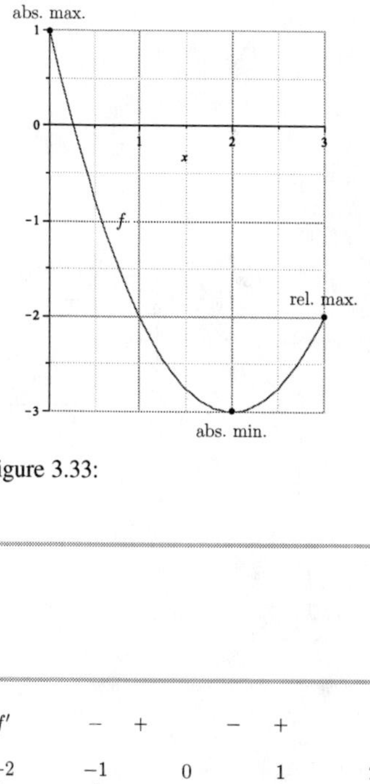

Figure 3.33:

2. (i) Find the critical numbers of f on $D_f = \mathbb{R}$:

$$f'(x) = 4x^3 - 2x = 4x\big(x^2 - \tfrac{1}{2}\big).$$

   Clearly, $f'(x) = 0$ is equivalent to $x = 0$ or $x = \pm\sqrt{\frac{1}{2}} = \pm\frac{1}{2}\sqrt{2}$. Since f' is defined on $D_f = \mathbb{R}$ and $\mathbb{R}$ does not have any boundary, the critical numbers of f are $\{-\frac{1}{2}\sqrt{2}, 0, \frac{1}{2}\sqrt{2}\}$.

   (ii) Determine the type of the critical numbers: We will apply the First Derivative Test. Pick sample points in each subinterval determined by the critical numbers. We choose $-1, -\frac{1}{2}, \frac{1}{2}, 1$.

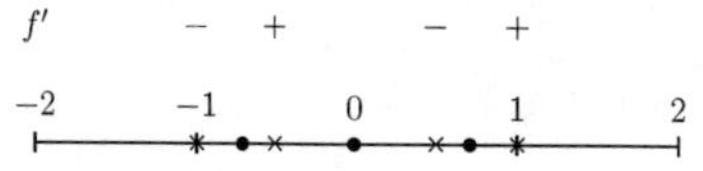

Figure 3.34:

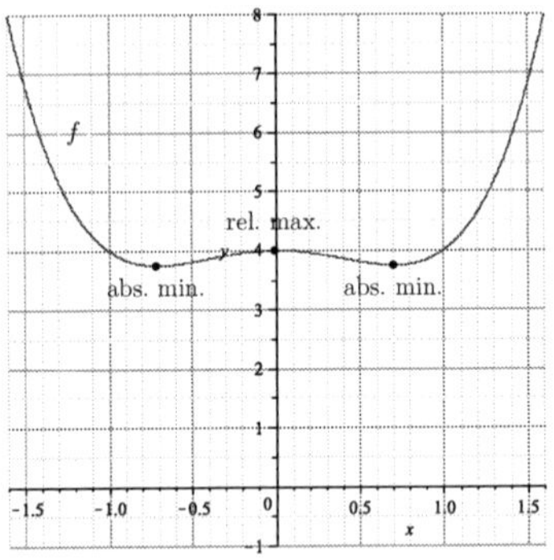

Figure 3.35:

Evaluating f' at these sample points (we only need the sign), we find that $f'(-1) < 0$, $f'(-\frac{1}{2}) > 0$, $f'(\frac{1}{2}) < 0$ and $f'(1) > 0$. By the Intermediate-Value Theorem, f' keeps its sign on the subintervals. Thus, by The First Derivative Test, f has relative minimum at $-\frac{1}{2}\sqrt{2}$, a relative maximum at 0 and a relative minimum at $\frac{1}{2}\sqrt{2}$ (Note that f is even).

(iii) Evaluate f at each extremum and compare the values: $f(\pm\frac{1}{2}\sqrt{2}) = \frac{15}{4}$ and $f(0) = 4$. Hence, $(0, 4)$ is a relative maximum and $(\pm\frac{1}{2}\sqrt{2}, \frac{15}{4})$ relative minima of$f$. In fact, these minima are absolute minima of f. Note that f is even and f is decreasing on the interval $(0, \frac{1}{2}\sqrt{2})$ and increasing on $(\frac{1}{2}\sqrt{2}, \infty)$, i.e. $f(x) \geq f(\frac{1}{2}\sqrt{2}) = \frac{15}{4}$ for all $x \in (0, \infty)$.

3. (i) Find the critical numbers of f on $D_f = \mathbb{R}$:

$$\begin{aligned} f'(x) &= 3x^2(4-x)^2 - 2x^3(4-x) \\ &= x^2(4-x)(3(4-x) - 2x) \\ &= 5x^2(x-4)(x - \tfrac{12}{5}) \end{aligned}$$

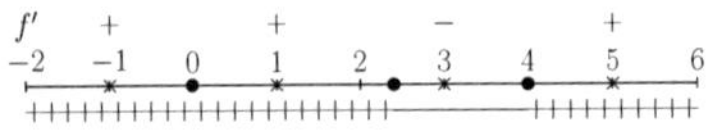

Figure 3.36:

Clearly, $f'(x) = 0$ is equivalent to $x = 0$ or $x = 4$ or $x\frac{12}{5}$. Since f' is defined for all $x \in \mathbb{R}$, the critical numbers of f are: $\{0, \frac{12}{5}, 4\}$.

(ii) Determine the type of the critical numbers: We will use the First Derivative Test. Pick sample points in each of the subintervals determined by the critical numbers. We choose -1, 1, 3 and 5.

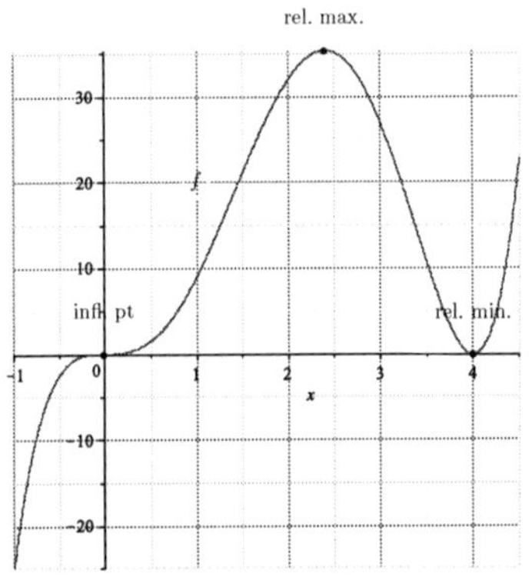

Figure 3.37:

Evaluating f' at those sample points (we only need the sign!), we find that $f'(-1) > 0$, $f'(1) > 0$, $f'(3) < 0$ and $f'(5) > 0$. By the Intermediate-Value Theorem, f' keeps its sign on the subintervals. Thus, by the First Derivative Test, f has an inflection point at 0, a relative maximum at $\frac{12}{5}$ and a relative minimum at 4.

(iii) Evaluate f at each extremum and compare the values: $f(\frac{12}{5}) = \frac{110\,592}{3\,125}$, $f(4) = 0$. Hence f has a relative maximum at $(\frac{12}{5}, \frac{110\,592}{3\,125})$ and a relative minimum at $(4, 0)$, but does not have any absolute extrema! Note that f does not satisfy the assumptions of the Extreme Value Theorem, since its domain is not a closed and bounded interval!

7. (i) Find the critical numbers of f on its domain: The domain of f is $D_f := [-2, 2]$. We first compute the derivative of f:

$$\begin{aligned} f'(x) &= \sqrt{4-x^2} - \frac{x^2}{\sqrt{4-x^2}} \\ &= \frac{-2(x^2-2)}{\sqrt{4-x^2}}. \end{aligned}$$

Note that f' is *not* defined at $-2, 2 \in D_f$, i.e. $D_{f'} = (-2, 2)$! Clearly, $f'(x) = 0$ if and only if $x = \pm\sqrt{2}$. Hence, the critical points of f on its domain D_f are $\{-2, -\sqrt{2}, \sqrt{2}, 2\}$.

(ii) Determine the type of each critical number: We will use the First Derivative Test and pick a sample point in each of the subintervals of $(-2, 2)$ determined by the critical numbers. We choose -1.7, 0 and 1.7. Evaluating f' at these points (we only need the sign), we find that $f'(\pm 1.7) < 0$ and $f'(0) > 0$. By the Intermediate-Value Theorem, f' keeps its sign on the whole subintervals.

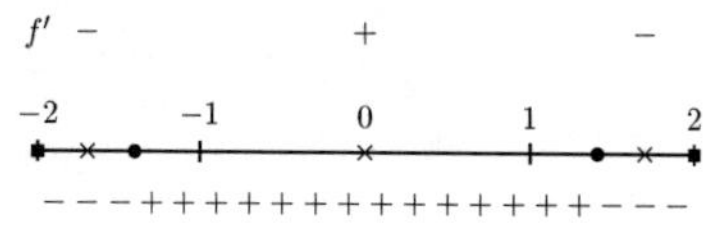

Figure 3.38:

Thus, by the First Derivative Test, f has a relative maximum at -2, a relative minimum at $-\sqrt{2}$, a relative maximum at $\sqrt{2}$ and a relative minimum at 2.

(iii) Evaluate f at each critical number and compare the values: $f(\pm 2) = 0$, $f(\pm\sqrt{2}) = \pm 2$, note that f is odd!. Hence, f has a relative maximum at $(-2, 0)$, an absolute minimum at $(-\sqrt{2}, -2)$, an absolute maximum at $(\sqrt{2}, 2)$ and a relative minimum at $(2, 0)$.

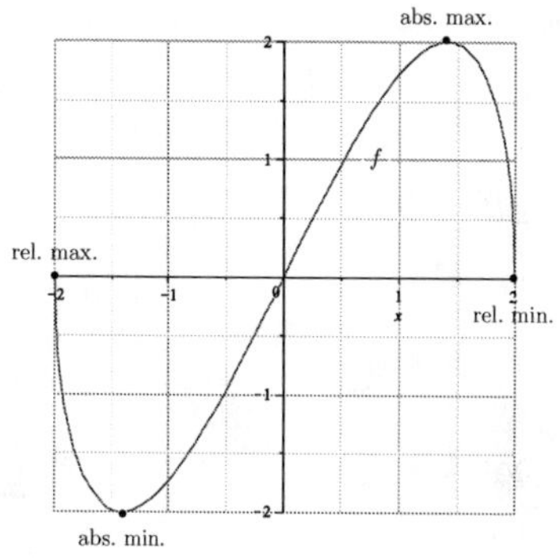

Figure 3.39:

8. (i) Find the critical numbers of f on its domain $[0, \pi]$: Differentiating f we obtain

$$\begin{aligned} f'(x) &= 2\cos^3 x + 3\sin x \\ &= 6\cos x\big[-\sin x\cos x + \tfrac{1}{2}\big] \\ &= -6\cos x\big[\tfrac{1}{2}\sin 2x - \tfrac{1}{2}\big] \\ &= -3\cos x\big[\sin 2x - 1\big], \end{aligned}$$

which is defined on $[0, \pi]$. Clearly $f'(x) = 0$ is equivalent to $\cos x = 0$ or $\sin 2x = 1$, i.e. $x = \frac{\pi}{2}$ or $x = \frac{\pi}{4}$. Hence, the critical numbers of f on $[0, \pi]$ are $\{0, \frac{\pi}{4}, \frac{\pi}{2}, \pi\}$.

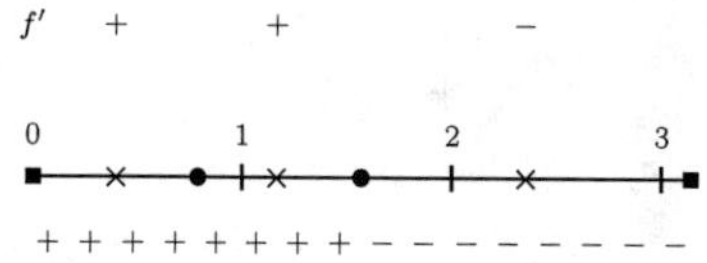

Figure 3.40:

(ii) Determine the type of each critical number: We will use the First Derivative Test and pick a sample point in each of the subintervals of $(0, \pi)$ determined by the critical numbers. We choose $\frac{1}{8}\pi$, $\frac{3}{8}\pi$ and $\frac{3}{4}\pi$. Evaluating f' at these points (we only need the sign), we find that $f'(\frac{1}{8}\pi) > 0$, $f'(\frac{3}{8}\pi) > 0$ and $f'(\frac{3}{4}\pi) < 0$.

By the Intermediate-Value Theorem, f' keeps its sign on the whole subintervals. Thus, by the First Derivative Test, f has a relative minimum at 0, an inflection point at $\frac{\pi}{8}$, a relative maximum at $\frac{\pi}{2}$ and a relative minimum at π.

(iii) Evaluate f at each critical number (representing an extremum) and compare the values: $f(0) = 2$, $f(\frac{\pi}{2}) = 3$ and $f(\pi) = -2$. Hence f has relative minimum at $(0, 2)$, an absolute maximum at $(\frac{\pi}{2}, 3)$ and an absolute minimum at $(\pi, -2)$.

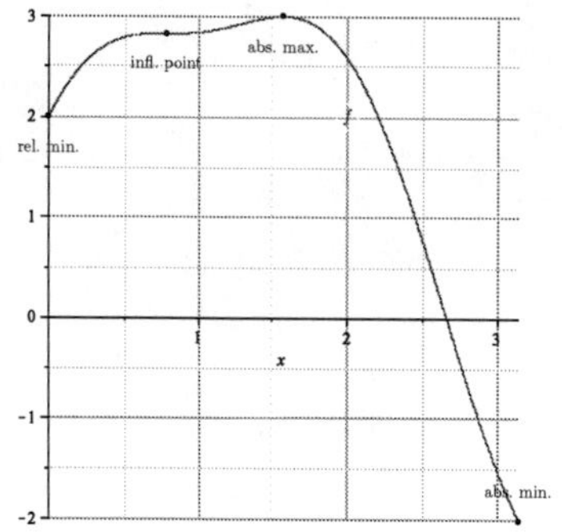

Figure 3.41:

9. (i) Find the critical numbers of f on its domain $D_f = [-3, 4)$: We can

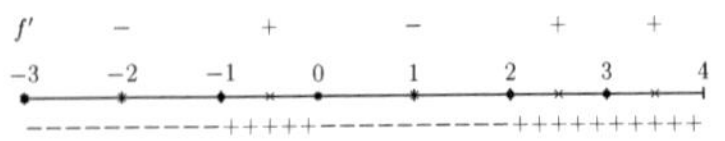

Figure 3.42:

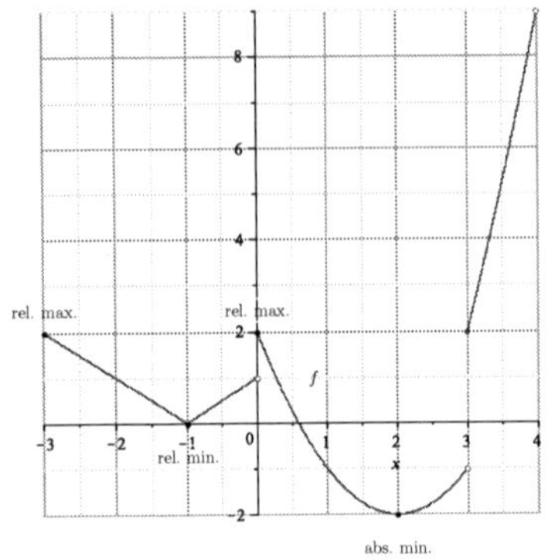

Figure 3.43:

rewrite f as

$$f(x) = \begin{cases} -(x+1) & -3 \leq x < -1 \\ x+1 & -1 \leq x < 0 \\ x^2 - 4x + 2 & 0 \leq x < 3 \\ 2x - 7 & 3 \leq x < 4. \end{cases}$$

Since

$$1 = \lim_{x \to 0^-} f(x) \neq \lim_{x \to 0^+} f(x) = 2$$

and

$$-1 = \lim_{x \to 3^-} f(x) \neq \lim_{x \to 3^+} f(x) = 2,$$

f is not continuous at 0 and 3 and therefore also not differentiable at 0 and 3. Moreover, since left- and right-sided derivative of f at -1 are distinct, f is also not differentiable at -1. Its derivative is

$$f'(x) = \begin{cases} -1 & -3 \leq x < -1 \\ 1 & -1 < x < 0 \\ 2x - 4 & 0 < x < 3 \\ 2 & 3 < x < 4. \end{cases}$$

with $D_{f'} = [-3, -1) \cup (-1, 0) \cup (0, 3) \cup (3, 4)$ as its domain. Clearly $f'(x) = 0$ is equivalant to $x = 2$. Thus, the critical numbers of f on $[-3, 4)$ are $\{-3, -1, 0, 2, 3\}$.

(ii) Determine the type of each critical number: We will use the First Derivative Test and pick a sample point in each of the subintervals of $[-3, 4)$ determined by the critical numbers. We choose -2, $-\frac{1}{2}$, 1, $\frac{5}{2}$ and $\frac{7}{2}$. Evaluating f' at those points (we only need the sign), we find that $f'(-2) < 0$, $f'(-\frac{1}{2}) > 0$, $f'(1) < 0$, $f'(\frac{5}{2}) > 0$ and $f'(\frac{7}{2}) > 0$.

By the Intermediate-Value Theorem, f' keeps its sign on the whole subintervals. Thus, by the First Derivative Test, f has a relative maximum at -3, a relative minimum at -1, a relative maximum at 0 and a relative minimum at 2. Hence,

(iii) Evaluate f at the critical numbers (representing extremma): $f(-3) = 2$, $f(-1) = 0$, $f(0) = 2$, $f(2) = -2$. Hence, f has a relative maximum at $(-3, 2)$, a relative minimum at $(-1, 0)$, a relative maximum at $(0, 2)$ and an absolute minimum at $(2, -2)$, however no absolute maximum. Note that f does not satisfy the assumption of the Extreme Value Theorem!

3.5.15 Concavity, Inflection Points

Decribe the concavity of the graph of f and find the points of inflections:

5. (i) Find the critical numbers of f': Clearly, $f'(x) = \cos(x)$, $f''(x) = -\sin(x)$. f'' is defined on $\mathbb{R}$. $f''(x) = 0$ is equivalent to $\sin(x) = 0$, which in turn is equivalent to $x = n\pi$ for all $n \in \mathbb{Z}$.

   (ii) Determine the sign of f'' on each subinterval determined by the critical numbers of f': We pick the sample points $\frac{2n+1}{2}\pi$ for $n \in \mathbb{Z}$. By inspection, we find that $f''\left(\frac{m}{2}\pi\right) < 0$ if and only if $m \equiv 1 \mod 4$, and $f''\left(\frac{m}{2}\pi\right) > 0$ if and only if $m \equiv 3 \mod 4$, in other words, $f''\left(\frac{m}{2}\pi\right) < 0$ if m has remainder 1 by division by 4 and $f''\left(\frac{m}{2}\pi\right) > 0$ if m has remainder 3 by division by 4.

   By the Intermediate Value Theorem, we can conclude that f'' keeps its sign on each of those subinterval. Hence,

   $$f''(x) > 0 \Leftrightarrow x \in \big((2n-1)\pi, 2n\pi\big),\ n \in \mathbb{Z}$$
   $$f''(x) < 0 \Leftrightarrow x \in \big(2n\pi, (2n+1)\pi\big),\ n \in \mathbb{Z},$$

   and thus f is concave up on

   $$\bigcup_{n\in\mathbb{Z}} \Big((2n-1)\pi, 2n\pi\Big),$$

   and concave down on

   $$\bigcup_{n\in\mathbb{Z}} \Big(2n\pi, (2n+1)\pi\Big).$$

   Each of the points $(n\pi, n\pi)$, for $n \in \mathbb{Z}$ is an inflection point of f.

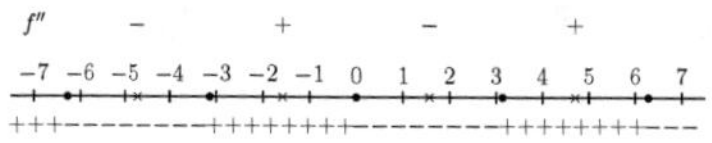

Figure 3.44:

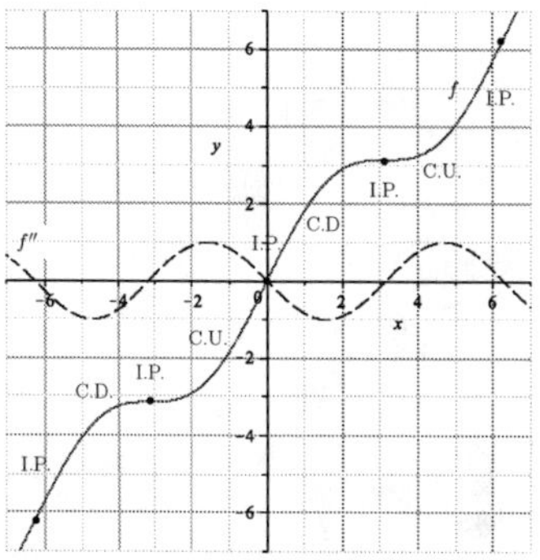

Figure 3.45:

6. (i) Find the critical numbers of f': Clearly,

   $$f'(x) = \begin{cases} 3x^2, & x < 1 \\ 3, & x \geq 1 \end{cases}$$

   Note that f is differentiable at $x = 1$. However f'' does *not* exist at $x = 1$:

   $$f''(x) = \begin{cases} 6x, & x < 1 \\ 0, & x > 1. \end{cases}$$

   $f''(x) = 0$ is equivalent to $x = 0$ or $x > 1$. Thus, the critical numbers of f' are $x = 0$ and $x = 1$, since $f''(1)$ does not exist. We do not have to consider the "non-discrete" zeroes.

   (ii) Determine the sign of f'' on each subinterval determined by the critical numbers of f': Since f'' is linear, we immediately see that $f''(x) < 0$ for $x \in (-\infty, 0)$, $f''(x) > 0$ for $x \in (0, 1)$ and $f''(x) = 0$ for $x \in (1, \infty)$. Hence, f is concave down on $(-\infty, 0)$, concave up on $(0, 1)$, linear on $(1, \infty)$ and has an inflection point at $(0, 0)$.

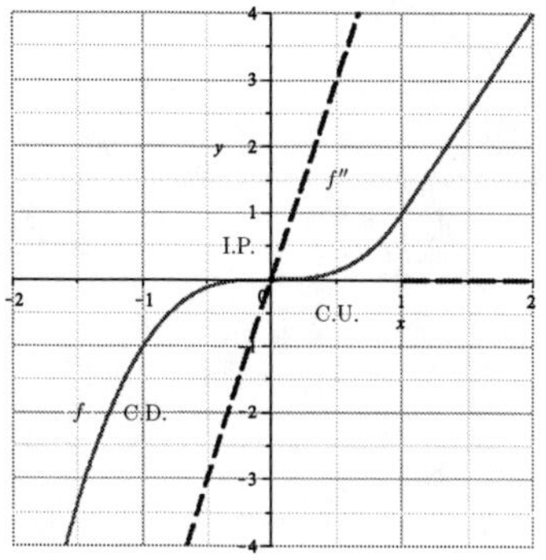

Figure 3.46:

3.5.16 Monotonicity

Find the intervals on which f increases/decreases:

3.5.17 Graph Sketching

Sketch the graph of the function f:

3.5.18 Maximum/Minimum Problems

6. Suppose the conical cup has radius r and height h and volume V. We would like to minimize its (surface) area A and will show that A attains its absolute minimum at $h = \sqrt{2}r$.

   Recall that

   $$V = \tfrac{1}{3}\pi r^2 h$$

   and thus

   $$h = \frac{3V}{\pi r^2}.$$

   Moreover,

   $$\begin{aligned} A = \pi r\sqrt{r^2 + h^2} &= \pi\left(r^2 + \frac{9V^2}{\pi^2 r^4}\right) \\ &= \left(\frac{\pi^2 r^6 + 9V^2}{r^2}\right). \end{aligned}$$

   Note that A gives the surface area of the conical cup of radius $r > 0$ with fixed volume V. In order to find its minimum, we first compute its derivative

   $$\begin{aligned} A'(r) &= \frac{\dfrac{6\pi^2 r^7 - 2r(\pi^2 r^6 + 9V^2)}{r^4}}{2\sqrt{\dfrac{\pi^2 r^6 + 9V^2}{r^2}}} \\ &= \frac{4\pi^2 r^7 - 18rV^2}{2r^4\sqrt{\dfrac{\pi^2 r^6 + 9V^2}{r^2}}} \\ &= \frac{2\pi^2 r^6 - 9V^2}{r^3\sqrt{\dfrac{\pi^2 r^6 + 9V^2}{r^2}}} \end{aligned}$$

It can be easily seen that $A'(r) = 0$ if and only if $r^3 = \frac{3V}{\sqrt{2}\pi}$ and that A has an absolute minimum at this point. Hence, since $3V = h\pi r^2$,

$$r^3 = \frac{3V}{\sqrt{2}\pi} = \frac{h\pi r^2}{\sqrt{2}, \pi} = \frac{hr^2}{\sqrt{2}}$$

which is equivalent to $r\sqrt{2} = h$, as claimed.

4

INTEGRAL CALCULUS

4.1 Definitions and Results

The Riemann integral that is reviewed in this chapter depends explicitly on the order structure of the real numbers. We therefore begin integration of real-valued functions that are defined on an interval.

4.1.1 Definition of the Riemann Integral

Definition 4.1 (Supremum, Infimum)
If $A \subseteq \mathbb{R}$ is a subset of the set of real numbers $\mathbb{R}$, then we define $\sup(A)$ to be the least upper bound of A if A is bounded above and ∞ if it is not bounded above. Similarly, we define $\inf(A)$ to be the greatest lower bound of A if A is bounded below and $-\infty$ if it is not bounded below.

Definition 4.2 (Partition)
Let $[a, b]$ be a closed interval. A subset $P := \{x_0, x_1, \ldots, x_n\}$ of $[a, b]$ is called a *partition* of $[a, b]$ if

$$a = x_0 < x_1 < \cdots < x_n = b.$$

Definition 4.3 (Upper Sums, Lower Sums)
Suppose the function f is bounded on the (closed and bounded) interval $[a, b]$ and $P := \{x_0, x_1, \ldots, x_n\}$ is a partition of $[a, b]$. Then we set

$$M_i(f) := \sup\{f(x) \mid x \in [x_{i-1}, x_i]\}$$
$$m_i(f) := \inf\{f(x) \mid x \in [x_{i-1}, x_i]\},$$

and $\Delta x_i := x_i - x_{i-1}$, and call the sums

$$U(f, P) := \sum_{i=1}^{n} M_i(f)\Delta x_i$$
$$L(f, P) := \sum_{i=1}^{n} m_i(f)\Delta x_i$$

upper sum and *lower sum* of f with respect to P, respectively.

Lemma 4.1
Suppose P and Q are partitions of $[a, b]$. Then

1. $L(f, P) \leq U(f, Q)$.
2. *If $P \subseteq Q$ then $U(f, Q) \leq U(f, P)$ and $L(f, Q) \geq L(f, P)$.*

Definition 4.4 (Upper and Lower Integral)
Suppose the function f is bounded on the interval $[a, b]$. Then

$$\overline{\int_a^b} f(x)\,dx := \inf\{U(f, Q) \mid Q \text{ partition of } [a, b]\}$$
$$\underline{\int_a^b} f(x)\,dx := \sup\{L(f, Q) \mid Q \text{ partition of } [a, b]\},$$

are called the *upper* and *lower Riemann integral,* respectively.

Remark 4.1
From Lemma 4.1 follows immediately that

$$\underline{\int_a^b} f(x)\,dx \leq \overline{\int_a^b} f(x)\,dx$$

Definition 4.5 (Riemann Integral)
Suppose the function f is bounded on $[a, b]$. We say that f is *Riemann integrable* or simply *integrable* if $\underline{\int}_a^b f(x)\,dx = \overline{\int}_a^b f(x)\,dx$. In this case we define

$$\int_a^b f(x)\,dx := \underline{\int}_a^b f(x)\,dx = \overline{\int}_a^b f(x)\,dx,$$

which is called the *definite (Riemann) integral* or simply the *integral* of f from a to b. Moreover, we define

$$\int_a^a f(x)\,dx := 0$$
$$\int_b^a f(x)\,dx := -\int_a^b f(x)\,dx.$$

Figure 4.1:
Bernhard Riemann
(1826–1866)

Theorem 4.1 (Integrability Criterion)
A bounded function f on $[a, b]$ is Riemann integrable if and only if for all $\varepsilon > 0$ there exists a partition P of $[a, b]$ such that

$$U(f, P) - L(f, P) < \varepsilon.$$

Theorem 4.2 (Sufficient Conditions for Integrability)

1. *If f is (piecewise) monotonic (either increasing or decreasing) on $[a, b]$ then f is Riemann integrable.*
2. *If f is (piecewise) continuous on $[a, b]$ then f is Riemann integrable.*

Theorem 4.3 (Approximation by Riemann Sums)
Suppose f is (Riemann) integrable on $[a, b]$. Then for any $\varepsilon > 0$ there exists a partition $P = \{x_0, \ldots, x_n\}$ of $[a, b]$ such that for any choice of $\xi_i \in [x_{i-1}, x_i]$

$$\left| \int_a^b f(x)\,dx - \sum_{i=1}^{n} f(\xi_i)\Delta x_i \right| < \varepsilon.$$

The sum $\sum_{i=1}^{n} f(\xi_i)\Delta x_i$ is called a Riemann sum *of f with respect to the partition P.*
In particular: If f is (Riemann) integrable on $[a, b]$ then

$$\int_a^b f(x)\,dx = \lim_{n\to\infty} \sum_{j=1}^{n} f(x_j)\,\Delta x,$$

where

$$\Delta x = \frac{b-a}{n}, \quad \text{and} \quad x_j = a + j\,\Delta x,$$

for $j = 0, \ldots, n$.

4.1.2 Estimating Definite Integrals

Definition 4.6
Given an Riemann integrable function f defined on $[a, b]$, choose $n \in \mathbb{N}$ and subdivide $[a, b]$ in n subintervals of equal length

$$\Delta x := \frac{b-a}{n}.$$

Let

$$x_i := a + i\,\Delta x.$$

for $i = 0, 1, \ldots, n$. The partition $P := \{x_0, \ldots, x_n\}$ is called an

Theorem 4.4 (Left-Hand and Right-Hand Sums)
For $n \in \mathbb{N}$, let

$$\mathrm{LHS}_f(n) := \sum_{i=0}^{n-1} f(x_i)\Delta x$$
$$\mathrm{RHS}_f(n) := \sum_{i=1}^{n} f(x_i)\Delta x.$$

If the function f is differentiable on $[a,b]$ and $|f'(x)| \leq B$ for all $x \in [a,b]$ and some $B \in \mathbb{R}$, then

$$\left| \int_a^b f(x)\,dx - \mathrm{LHS}_f(n) \right| \leq \frac{B(b-a)^2}{2n}$$
$$\left| \int_a^b f(x)\,dx - \mathrm{RHS}_f(n) \right| \leq \frac{B(b-a)^2}{2n}.$$

Theorem 4.5 (Trapezoidal Estimate)
For $n \in \mathbb{N}$, let

$$\mathrm{TRAP}_f(n) := \frac{1}{2}\Big(\mathrm{LHS}_f(n) + \mathrm{RHS}_f(n)\Big).$$

If the function f is twice differentiable on $[a,b]$ and $|f''(x)| \leq B$ for all $x \in [a,b]$ for some $b \in \mathbb{R}$, then

$$\left| \int_a^b f(x)\,dx - \mathrm{TRAP}_f(n) \right| \leq \frac{B(b-a)^3}{12n^2}.$$

Theorem 4.6 (Midpoint Sum)
For $n \in \mathbb{N}$, let $\mu_i := \frac{1}{2}(x_{i-1} + x_i) = a + (i - \frac{1}{2})\Delta x$ denote the midpoint of the ith subinterval of $[a,b]$. Moreover, let

$$\mathrm{MPS}_f(n) := \sum_{i=1}^{n} f(\mu_i)\Delta x.$$

If the function is twice differentiable on $[a,b]$ and $|f''(x) \leq B$ for all $x \in [a,b]$ for some $b \in \mathbb{R}$. then

$$\left| \int_a^b f(x)\,dx - \mathrm{MPS}_f(n) \right| \leq \frac{B(b-a)^3}{24n^2}.$$

Figure 4.2:
Thomas Simpson
(1710–1761)

Theorem 4.7 (Simpson's Rule)
For $n \in \mathbb{N}$, let

$$\begin{aligned}\mathrm{SIMP}_f(2n) :&= \frac{1}{3}\Big(2 \cdot \mathrm{MPS}_f(n) + \mathrm{TRAP}_f(n)\Big) \\ &= \sum_{i=1}^{n}\Big(f(x_{i-1} + 4f(\mu_i) + f(x_i)\Big)\frac{\Delta x}{6}.\end{aligned}$$

If the function f is four times differentiable on $[a,b]$ and $|f^{(4)}(x)| \le B$ for all $x \in [a,b]$ for some $B \in \mathbb{R}$, then

$$\left|\int_a^b f(x)\,dx - \mathrm{SIMP}_f(n)\right| \le \frac{B(b-a)^5}{180n^4}.$$

4.1.3 Properties of the Definite Integral

Theorem 4.8 (Linearity)
Suppose f and g are integrable and $\alpha \in \mathbb{R}$, then

1. $\displaystyle\int_a^b \alpha f(x)\,dx = \alpha\int_a^b f(x)\,dx$
2. $\displaystyle\int_a^B \big(f(x) + g(x)\big)\,dx = \int_a^b f(x)\,dx + \int_a^b g(x)\,dx$

Theorem 4.9 (Interval Property)
Suppose f is integrable on the interval I and $a, b, c \in I$, then

$$\int_a^c f(x)\,dx = \int_a^b f(x)\,dx + \int_b^c f(x)\,dx.$$

Theorem 4.10 (Order Properties)

1. Positivity: *If the function f is integrable on $[a, b]$ and $f(x) \geq 0$ for all $x \in [a, b]$, then $\int_a^b f(x)\,dx \geq 0$.*
   In particular: If the functions f and g are integrable on $[a, b]$ and $f(x) \leq g(x)$ for all $x \in [a, b]$, then $\int_a^b f(x)\,dx \leq \int_a^b g(x)\,dx$.

2. Triangle Inequality: *If f is integrable on $[a, b]$, then $|f|$ is integrable on $[a, b]$ and*
$$\left| \int_a^b f(x)\,dx \right| \leq \int_a^b |f(x)|\,dx.$$

3. Estimate: *If f is integrable on $[a, b]$ and m the minimum and M the maximum value of f on $[a, b]$ then*
$$m(b-a) \leq \int_a^b f(x)\,dx \leq M(b-a).$$

Theorem 4.11 (Symmetry)
Suppose f is integrable on the symmetric interval $[-a, a]$. Then,

1. *if f is odd on $[-a, a]$, then*
$$\int_{-a}^a f(x)\,dx = 0;$$

2. *if f is even on $[-a, a]$, then*
$$\int_{-a}^a f(x)\,dx = 2\int_0^a f(x)\,dx.$$

Definition 4.7 (Average Value of a Function)
If f is integrable on $[a, b]$, then
$$\operatorname{av}_f := \frac{1}{b-a}\int_a^b f(x)\,dx$$
is the *average value* of f on the interval $[a, b]$.

Theorem 4.12 (Mean Value Theorem for Integrals)
If f is continuous on $[a, b]$, then there exists a number $\xi \in [a, b]$ such that
$$f(\xi) = \frac{1}{b-a}\int_a^b f(x)\,dx.$$

Theorem 4.13 (Second Mean Value Theorem for Integrals)
If f and g are continuous on $[a, b]$ and $g(x) \geq 0$ for all $x \in [a, b]$, then there exists a number $\xi \in [a, b]$ such that

$$\int_a^b f(x)g(x)\,dx = f(\xi)\int_a^b g(x)\,dx.$$

4.1.4 The Fundamental Theorem of Calculus

Definition 4.8 (Anti-Derivative, Primitive)
Let f be defined on the interval I. A function F is called an *anti-derivative* or *primitive* of the function f on I if $F'(x) = f(x)$ for all $x \in I$.

Definition 4.9 (Indefinite Integral)
Let f be a function defined on some interval I. Then the family of all anti-derivatives of f is called the *indefinite integral* of f and denoted by

$$\int f(x)\,dx$$

Thus, if F is any anti-derivative of f on I, then

$$\int f(x)\,dx = F(x) + C$$

for $C \in \mathbb{R}$.

Theorem 4.14 (Fundamental Theorem of Calculus)
Suppose f is continuous on the interval $[a, b]$. Then

1. *The function*
$$F(x) := \int_a^x f(t)\,dt$$
is differentiable on $[a, b]$ and
$$F'(x) = \frac{d}{dx}\left(\int_a^x f(t)\,dt\right) = f(x)$$
for all $x \in [a, b]$.

2. *If G is any anti-derivative of f on $[a, b]$, then*
$$\int_a^b f(x)\,dx = G(x)\Big|_a^b = G(b) - G(a).$$

Corollary 4.1
If f is continuous on $[a,b]$ and ψ_1 and ψ_2 are differentiable on $[a,b]$ then

$$G(x) := \int_{\psi_1(x)}^{\psi_2(x)} f(t)\,dt$$

ist differentiable on $[a,b]$ and

$$G'(x) = \frac{d}{dx}\left(\int_{\psi_1(x)}^{\psi_2(x)} f(t)\,dt\right) = f(\psi_2(x))\,\psi_2'(x) - f(\psi_1(x))\,\psi_1'(x).$$

4.1.5 Integrals of Elementary Functions

Theorem 4.15 (Integrals I)

1. Power Rule:

$$\int x^r\,dx = \begin{cases} \dfrac{x^{r+1}}{r+1} + C, & r \neq -1 \\ \ln|x| + C, & r = -1 \end{cases}$$

2. Generalized Power Rule:

$$\int (g(x))^r g'(x)\,dx = \begin{cases} \dfrac{(g(x))^{r+1}}{r+1} + C, & r \neq -1 \\ \ln|g(x)| + C, & r = -1 \end{cases}$$

3. $\displaystyle\int \cos x\,dx = \sin x + C$

$\displaystyle\int \sin x\,dx = -\cos x + C$

4. $\displaystyle\int \tan x\,dx = -\ln|\cos x| + C$

$\displaystyle\int \cot x\,dx = \ln|\sin x| + C$

5. $\displaystyle\int \sec x\,dx = \ln|\sec x + \tan x| + C$

$\displaystyle\int \csc x\,dx = -\ln|\csc x + \cot x| + C.$

6. $\displaystyle\int \sec^2\,dx = \tan x + C$

$\displaystyle\int \csc^2 x\,dx = -\cot x + C$

7. $\displaystyle\int \sec^3 x\,dx = \tfrac{1}{2}\sec x\tan x + \tfrac{1}{2}\ln|\sec x + \tan x| + C$

$\displaystyle\int \csc^3 x\,dx = -\tfrac{1}{2}\csc x\cot x - \tfrac{1}{2}\ln|\csc x + \cot x| + C$

8. $\displaystyle\int \sec x\tan x\,dx = \sec x + C$

$\displaystyle\int \csc x\cot x\,dx = -\csc x + C$

9. $\displaystyle\int e^x\,dx = e^x + C$

$\displaystyle\int a^x\,dx = \frac{a^x}{\ln a} + C$

Theorem 4.16 (Integrals II)

1. $\displaystyle\int \frac{1}{1+x^2}\,dx = \tan^{-1} x + C$

2. $\displaystyle\int \frac{1}{\sqrt{1-x^2}}\,dx = \sin^{-1} x + C$

3. $\displaystyle\int \frac{1}{x\sqrt{x^2-1}}\,dx = \sec^{-1} x + C$

4. $\displaystyle\int \cosh x\,dx = \sinh x + C$

   $\displaystyle\int \sinh x\,dx = \cosh x + C$

5. $\displaystyle\int \operatorname{sech}^2 x\,dx = \tanh x + C$

   $\displaystyle\int \operatorname{csch}^2 x\,dx = -\coth x + C$

6. $\displaystyle\int \operatorname{sech} x \tanh x\,dx = -\operatorname{sech} x + C$

   $\displaystyle\int \operatorname{csch} x \coth x\,dx = -\operatorname{csch} x + C$

7. $\displaystyle\int \cos^n x\,dx = \tfrac{1}{n}\cos^{n-1} x \sin x + \tfrac{n-1}{n}\int \cos^{n-2} x\,dx$

   $\displaystyle\int \sin^n x\,dx = -\tfrac{1}{n}\sin^{n-1} x \cos x + \tfrac{n-1}{n}\int \sin^{n-2} x\,dx$

8. $\displaystyle\int \tan^n x\,dx = \tfrac{1}{n-1}\tan^{n-1} x - \int \tan^{n-1} x\,dx$

   $\displaystyle\int \cot^n x\,dx = -\tfrac{1}{n-1}\cot^{n-1} x - \int \cot^{n-1} x\,dx$

9. $\displaystyle\int \sec^n x\,dx = \tfrac{1}{n-1}\sec^{n-2} x \tan x + \tfrac{n-2}{n-1}\int \sec^{n-2} x\,dx$

   $\displaystyle\int \csc^n x\,dx = -\tfrac{1}{n-1}\csc^{n-2} x \cot x + \tfrac{n-2}{n-1}\int \csc^{n-2} x\,dx$

4.1.6 Integration Rules

Theorem 4.17 (Linearity)

1. $\int \alpha f(x)\,dx = \alpha \int f(x)\,dx$

2. $\int \big(f(x) + g(x)\big)\,dx = \int f(x)\,dx + \int g(x)\,dx.$

Theorem 4.18 (Substitution Rule)

1. $\int_a^b f\big(g(x)\big)g'(x)\,dx = \int_{g(a)}^{g(b)} f(y)\,dy$

2. $\int f\big(g(x)\big)g'(x)\,dx = \int f(y)\,dy$

Remark 4.2 (Some Standard Substitutions)

1. *Trigonometric Substitutions:*

   If the integrand involves $\sqrt{a^2 - x^2}$, $\sqrt{x^2 - a^2}$ or $\sqrt{a^2 + x^2}$, substitute $x = a\sin\theta$, $x = a\sec\theta$, $x = a\tan\theta$, respectively.

2. *Rationalizing Substitutions:*

   If the integrand involves an expression of the form $\sqrt[n]{f(x)}$ substitute $y = \sqrt[n]{f(x)}$ or equivalently, $y^n = f(x)$.

   If the integrand is a rational expression in $\sin x$ and $\cos x$, substitute $y = \tan(x/2)$, which implies

$$\cos x = \frac{1 - y^2}{1 + y^2}, \qquad \sin x = \frac{2y}{1 + y^2}$$
$$dx = \frac{2\,dy}{1 + y^2}$$

Theorem 4.19 (Integration by Parts)

1. $\int_a^b f'(x)g(x)\,dx = f(x)g(x)\Big|_a^b - \int_a^b f(x)g'(x)\,dx$

2. $\int f'(x)g(x)\,dx = f(x)g(x) - \int f(x)g'(x)\,dx$

Remark 4.3 (Integration by Partial Fraction Decomposition)
If the integrand is a proper rational function of the form $p(x)/q(x)$ where p and q are polynomials (if it is not proper, first use long division) follow this procedure:

(STEP 1:) Factor q into a product of linear and irreducible (over $\mathbb{R}$) quadratic factors.

(STEP 2:) Decompose $p(x)/q(x)$ by replacing each linear factor of q which has the form $(ax+b)^k$ (with multiplicity $k \geq 1$) by

$$\frac{A_1}{ax+b} + \frac{A_2}{(ax+b)^2} + \cdots + \frac{A_k}{(ax+b)^k}$$

and each quadratic factor of q which has the form $(ax^2+bx+c)^k$ (with multiplicity $k \geq 1$) by

$$\frac{A_1x+B_1}{ax^2+bx+c} + \cdots + \frac{A_kx+B_k}{(ax^2+bx+c)^k}.$$

(STEP 3:) Compute the undetermined coefficients A_j, B_j in the numerators using "comparing coefficients", limit techniques etc.

(STEP 4:) Solve the resulting integrals which are of the form

$$\int \frac{A_j}{(ax+b)^j)}\, dx, \quad \int \frac{A_jx+b_j}{(ax^2+bx+c)^j}\, dx$$

4.1.7 Improper Integrals

Recall that the Riemann integral (see Defintion 4.5 on page 207) is defined for bounded functions defined on a closed and bounded interval. We will now consider some extensions:

Definition 4.10 (Integrals over $[a,\infty)$)
Let the function f be defined on the interval $[a,\infty)$ and suppose the definite integral

$$\int_a^b f(x)\, dx$$

exists for all $b \in [a,\infty)$. Then we define

$$\int_a^\infty f(x)\, dx := \lim_{b\to\infty} \int_a^b f(x)\, dx$$

provided the limit exist and call it the *improper integral of f over $[a,\infty)$.*

Example 4.1 (p-Integrals)
Let $p \in \mathbb{R}$, then

$$\int_1^\infty \frac{1}{x^p}\, dx = \begin{cases} \dfrac{1}{p-a}, & \text{if } p > 1 \\ \infty, & \text{else} \end{cases}$$

Definition 4.11 (Integrals over $(-\infty, b]$)
Let the function f be defined on the interval $(\infty, b]$ and suppose the definite integral

$$\int_a^b f(x)\,dx$$

exists for all $a \in (\infty, b]$. Then we define

$$\int_{-\infty}^b f(x)\,dx := \lim_{a \to -\infty} \int_a^b f(x)\,dx$$

provided the limit exist and call it the *improper integral of f over* $(-\infty, b]$.

Definition 4.12 (Integrals over $(-\infty, \infty)$)
Let the function f be defined on $\mathbb{R}$ and suppose the improper integrals

$$\int_{-\infty}^c f(x)\,dx \quad \text{and} \quad \int_c^\infty f(x)\,dx$$

exist for all $c \in \mathbb{R}$, then we define

$$\int_{-\infty}^\infty f(x)\,dx := \int_{-\infty}^c f(x)\,dx + \int_c^\infty f(x)\,dx$$

Definition 4.13 (Integrals over Unbounded Functions)
Let f be defined on the interval $[a, b)$ and suppose f is unbounded at b. Suppose the definite integral

$$\int_a^c f(x)\,dx$$

exists for all $c \in [a, b)$. Then we define

$$\int_a^b f(x)\,dx := \lim_{c \to b} \int_a^c f(x)\,dx$$

provided the limit exists and call it the *improper integral over* $[a, b]$.

Theorem 4.20 (Comparison Theorem)
Let $a \in \mathbb{R}$ and suppose that f and g are continuous functions on the interval $[a, \infty)$ with

$$0 \leq f(x) \leq g(x) \quad \text{for all } x \in [a, \infty)$$

Then

1. *If $\int_a^\infty g(x)\,dx$ converges, then $\int_a^\infty f(x)\,dx$ converges.*
2. *If $\int_a^\infty f(x)\,dx$ diverges, then $\int_a^\infty g(x)\,dx$ diverges.*

Remark 4.4
Note that a similar result can be formulated for the other variations of improper integrals.

Example 4.2
Determine whether the improper integral $\int_0^\infty e^{-x^2}\,dx$ converges.

SOLUTION: As the antiderivative of e^{-x^2} is not an elementary function, we cannot evaluate the integral directly. Consider

$$\int_0^\infty e^{-x^2}\,dx = \int_0^1 e^{-x^2}\,dx + \int_1^\infty e^{-x^2}\,dx$$

Clearly, the first integral on the right-hand side is an ordinary definite integral. Therefore, we only have to determine whether the second integral on the right-hand side converges. However, since $-x^2 \leq -x$ for all $x \geq 1$ and exp is order preserving (increasing), it follows that $e^{-x^2} \leq e^{-x}$ for all $x \geq 1$. Note that

$$\int_1^\infty e^{-x}\,dx = \lim_{b\to\infty} \int_1^b e^{-x}\,dx = \lim_{b\to\infty} \left(e^{-1} - e^{-b}\right) = e^{-1}$$

Hence, by the Comparison Theorem (Theorem 4.20), also $\int_0^\infty e^{-x^2}\,dx$ converges. ◁

4.1.8 Some Applications of Integration

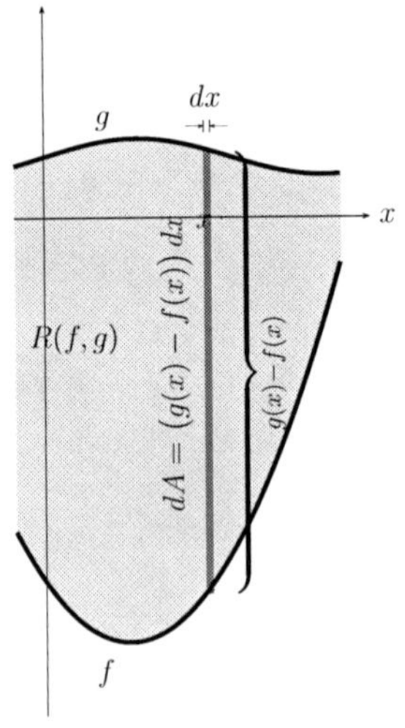

Figure 4.3:

Area between Two Curves

Remark 4.5
If f and g are continuous functions on $[a, b]$ such that $g(x) \leq f(x)$ for all $x \in [a, b]$, then the region

$$R(f, g) = \{(x, y) \mid x \in [a, b],\ g(x) \leq y \leq f(x)\}$$

is called the *region between f and g over* $[a, b]$. The *area differential* at $x \in [a, b]$ is given by

$$dA(x) = \big(f(x) - g(x)\big)\, dx$$

and the total area by

$$A = \int_a^b dA = \int_a^b \big(f(x) - g(x)\big)\, dx.$$

Example 4.3
Find the area between the curves $y = x$ and $y = x^2$ over the interval $[0, 1.5]$.

SOLUTION: Note the graphs of the functions $f(x) := x$ and $g(x) := x^2$ intersect at the points $(0, 0)$ and $(1, 1)$, more precisely

$$\begin{aligned} f(x) &\geq g(x) \quad \text{for } x \in [0, 1] \\ f(x) &\leq g(x) \quad \text{for } x \in [1, 1.5]. \end{aligned}$$

Therefore, we have to split the computation of the area between f and g over $[0, 1.5]$ into two parts:

$$dA_1 = \big(f(x) - g(x)\big)\, dx = (x - x^2)\, dx$$

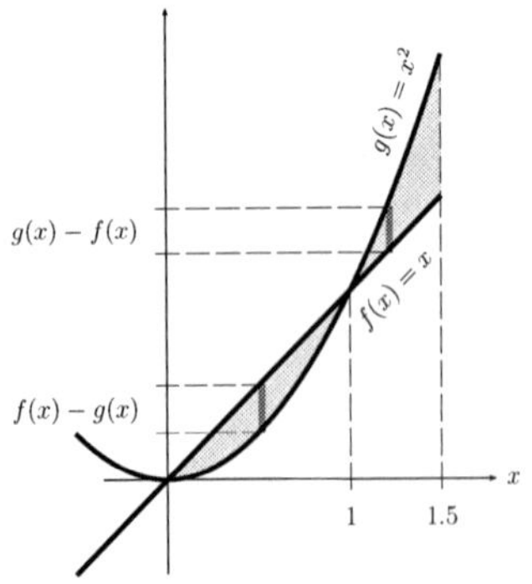

Figure 4.4:

for $x \in [0, 1]$ and

$$dA_2 = \big(g(x) - f(x)\big)\, dx = (x^2 - x)\, dx$$

for $x \in (1, 1.5]$. Here $dA_1(x)$ is the area of the xth slice of the shaded region for $x \in [0, 1]$, while $dA_2(x)$ is the area of the xth slice for $x \in (1, 1.5]$. The total area of the shaded region is given by the sum of the following two integrals

$$\begin{aligned} A &= \int_0^1 dA_1 + \int_1^{1.5} dA_2 \\ &= \int_0^1 (x - x^2)\, dx + \int_1^{1.5} (x^2 - x)\, dx \\ &= \left[\tfrac{1}{2}x^2 - \tfrac{1}{3}x^3\right]_0^1 + \left[\tfrac{1}{3}x^3 - \tfrac{1}{2}x^2\right]_1^{3/2} \\ &= \tfrac{1}{2} - \tfrac{1}{3} + \tfrac{9}{8} - \tfrac{9}{8} - \tfrac{1}{3} + \tfrac{1}{2} = \tfrac{1}{3}. \end{aligned}$$

◁

Volume by Slicing

Example 4.4
Compute the volume of a cylinder of radius r and height h.

SOLUTION: The following figure suggests that the volume of a cylinder is the sum” of the volumes $dV(x) = r^2\pi\, dx$ of the cross-sectional slices of thickness dx perpendicular to its axis:

$$\begin{aligned} V &= \int_a^b dV(x) = \int_a^b r^2\pi\, dx \\ &= \left[r^2\pi x\right]_a^b = r^2\pi(b-a). \end{aligned}$$

◁

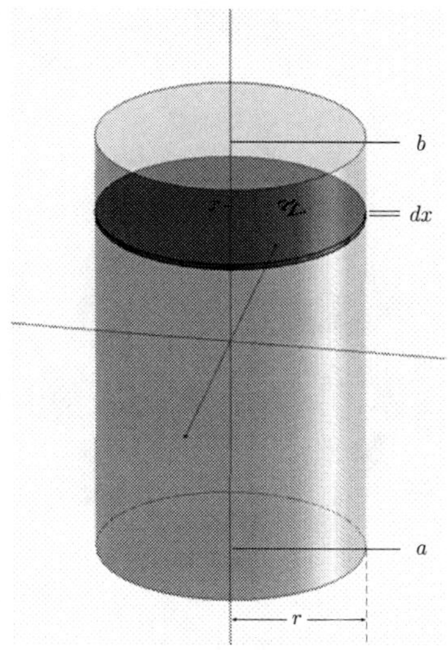

Figure 4.5: ‘
‘

Theorem 4.21 (Cavalieri's Principle)
Given a solid Ω with continuous cross-sections $A(x)$ perpendicular to a coordinate axis whose points are labeled by x. Then the volume differential *of Ω at the point x is given by*

$$dV(x) = A(x)\, dx$$

and if $a < b$ are points on the axis, then the volume of Ω between a and b is given by

$$V = \int_a^b dA = \int_a^b A(x)\, dx.$$

Example 4.5
Let $a, b \in \mathbb{R}$ and $a, b > 0$. The base of a solid is the region bounded by the ellipse defined by

$$\frac{x^2}{a^2} + \frac{y^2}{b^2} = 1.$$

Find the volume of the solid provided that each cross-section perpendicular to the x-axis is a square. See Figure 4.14.

SOLUTION: We only need to determine the area $A(x)$ of the x-th cross-section of the solid for arbitrary $x \in [-a, a]$. By assumption, each such cross-section is a square of base $b(x)$ which is given by $b(x) = 2b/a\sqrt{a^2 - x^2}$ Thus

$$A(x) = b(x)^2 = \frac{4b^2(a^2 - x^2)}{a^2}$$

and therefore

$$dV(x) = A(x)\, dx = \frac{4b^2(a^2 - x^2)}{a^2}\, dx.$$

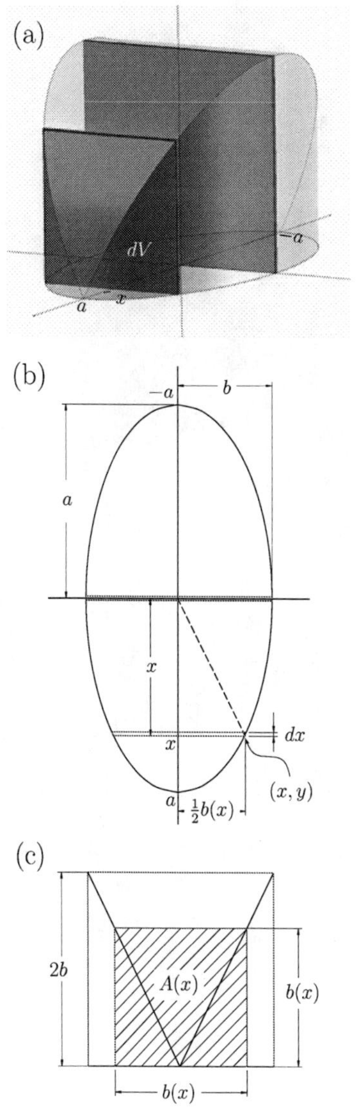

Figure 4.6:

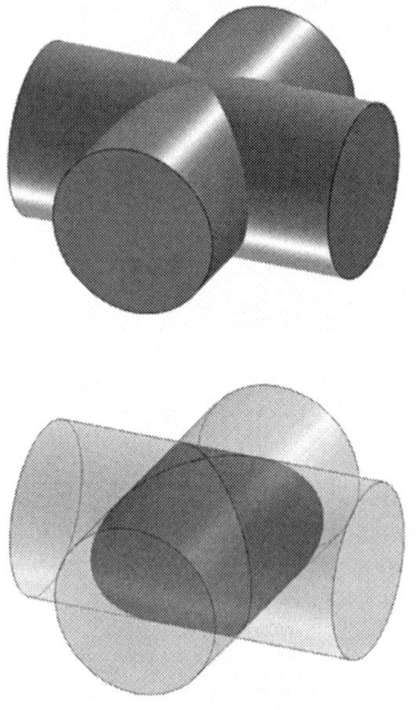

Figure 4.7:

The total volume of the solid is the integral

$$\begin{aligned} V &= \int_{-a}^{a} dV = \frac{4b^2}{a^2} \int_{-a}^{a} (a^2 - x^2)\, dx \\ &= \frac{4b^2}{a^2} \left(a^2 x - \tfrac{1}{3} x^3 \right) \Big|_{-a}^{a} \\ &= \frac{4b^2}{a^2} \left(2a^3 - \tfrac{2}{3} a^3 \right) = \frac{16}{3}\, b^2 a. \end{aligned}$$

◁

Example 4.6
Two cylinders of radius r intersect at right angles, their axis meeting. Find the volume of the intersection. See the figures below.

SOLUTION: To assist our intuition, we visualize the constellation of the two cylinders and their intersection. Note that the cross-section parallel to the plane determined by the axis of the two cylinders are all squares. show how the base $b(x)$ of the x-th square can be computed. Clearly,

$$b(x) = 2\sqrt{r^2 - x^2}$$

and therefore the volume of the xth slice is given by

$$dV(x) = A(x)\, dx = 4\big(r^2 - x^2\big)\, dx$$

for $x \in [-r, r]$. The total volume of the intersection of the two cylinders is thus the integral

$$\begin{aligned} V &= \int_{-r}^{r} dV(x) = 4 \int_{-r}^{r} (r^2 - x^2)\, dx \\ &= 8 \int_{0}^{r} (r^2 - x^2)\, dx \\ &= 8 \left(r^2 x - \tfrac{1}{3} x^3 \right) \Big|_{0}^{r} = \frac{16}{3}\, r^3. \end{aligned}$$

◁

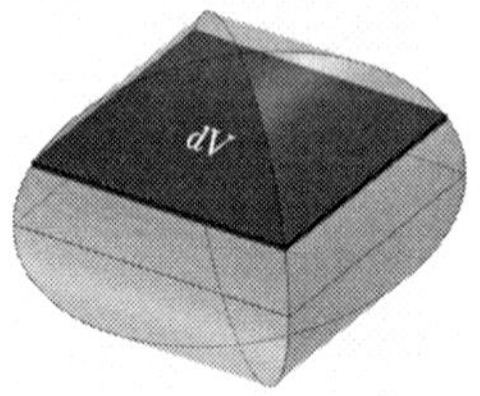

(a)

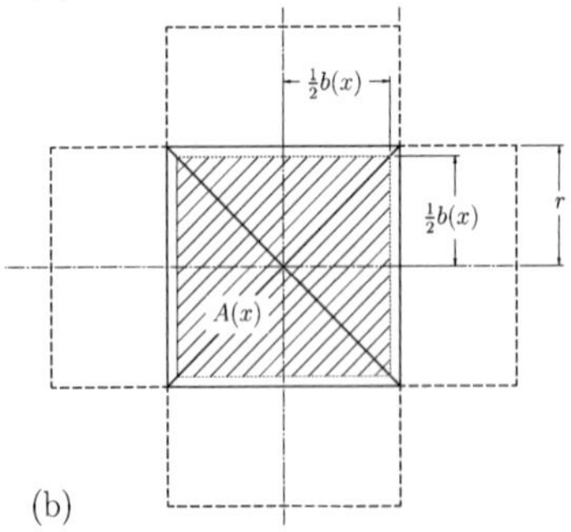

(b)

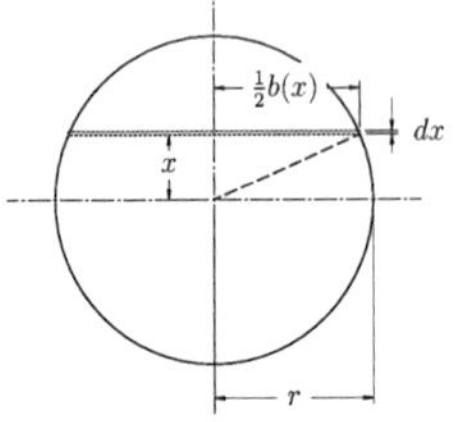

(c)

Figure 4.8:

Remark 4.6 (Solids of Revolution)
An important special case of solids with known cross-sections are the so-called "solids of revolution". Consider a region R in the plane and a line l that does not intersect the interior of R. A *solid of revolution* is generated if we revolve the region R through a full revolution in space about the line ℓ as axis. See the figure below depicting a torus being generated by revolving the highlighted circular region R about the axis ℓ.

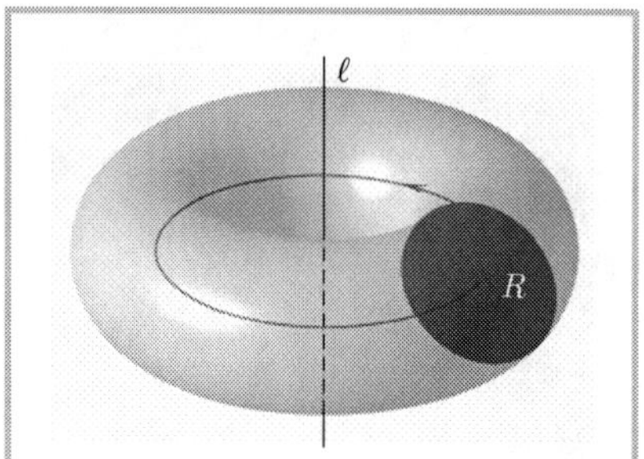

Figure 4.9:

Example 4.7
Find the volume of the solid generated by revolving the region R between $y = x^2$ and $y = 2x$ located in the I-th quadrant about the y-axis.

SOLUTION:

According to Cavalieri's Principle, it suffices to compute the area $A(y)$ of the y-th cross-section of the solid perpendicular to the y-axis constituting the axis of revolution ℓ in this case. Note that we have to choose y as the *independent variable,* since the regions R is revolved about the y-axis! In this case, where the curves bounding the region R are given by equations, we only have to solve these equations for the variable x. Figure 4.20 shows that each such cross-section is an annulus of inner radius $x = g(y) := \frac{1}{2}y$ and and outer radius $x = f(y) := \sqrt{y}$ with $y \in [0, 4]$. Hence, the solid has cross-sectional area

$$A(y) = \pi\left(\sqrt{y}\right)^2 - \pi\left(\tfrac{1}{2}y\right)^2 = \pi\left(y - \tfrac{1}{4}y^2\right)$$

and cross-sectional volume (volume of the y-th slice of the solid)

$$dV(x) = A(y)\,dy = \pi\left(y - \tfrac{1}{4}y^2\right)dy.$$

The total volume of the solid is therefore the integral

$$V = \int_0^4 dV(y) = \pi \int_0^4 \left(y - \tfrac{1}{4}y^2\right)dy$$
$$= \pi\left(\tfrac{1}{2}y^2 - \tfrac{1}{12}y^3\right)\Bigg|_0^4 = \tfrac{8}{3}\pi.$$

◁

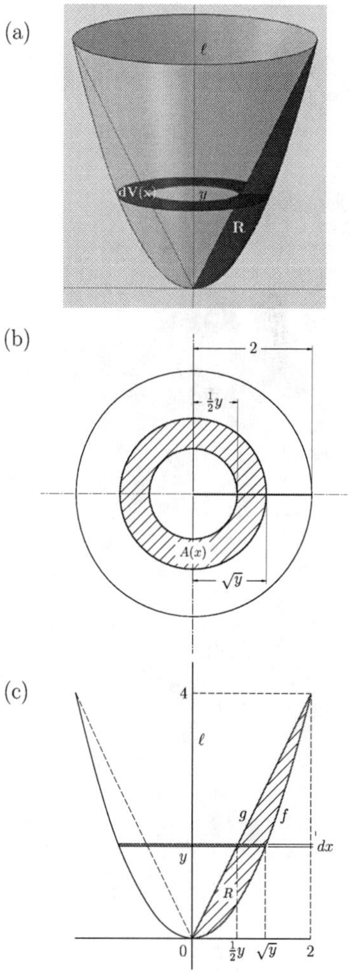

Figure 4.10:

Remark 4.7 (Disk/Washer Method)
Let $c \in \mathbb{R}$ and f and g continuous on $[a, b]$ such that $c \leq g(x) \leq f(x)$ for all $x \in [a, b]$. Moreover, let Ω denote the solid generated by revolving the region between f and g over $[a, b]$ about the axis $y = c$. Then the volume differential of Ω at the point $x \in [a, b]$ is given by

$$dV(x) = \pi\Big[\big(f(x) - c\big)^2 - \big(g(x) - c\big)^2\Big]\,dx$$

Figure 4.11:
Bonaventura Francesco Cavalieri (1598–1647)

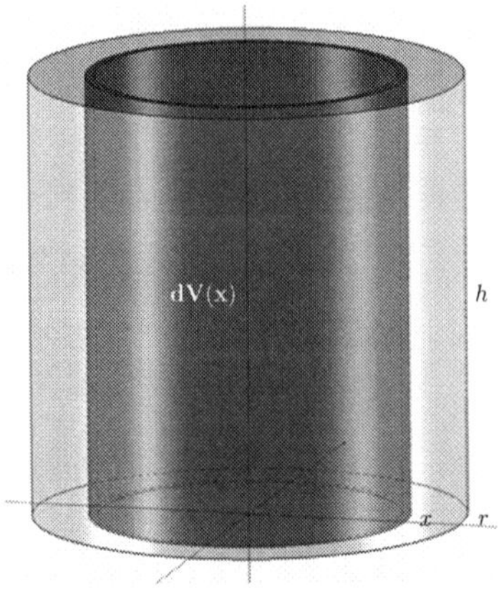

Figure 4.12:

and the total volume of Ω by

$$\begin{aligned} V &= \int_a^b dV \\ &= \pi \int_a^b \left[\big(f(x) - c\big)^2 - \big(g(x) - c\big)^2 \right] dx \end{aligned}$$

Volumes by Shells

Remark 4.8 (Shell Method)
Consider a solids of revolution where the axis of revolution is parallel (or equal) to the ordinate (the axis of the dependent variable) of the functions defining the bounding curves of the region R. Here we can use a "radial version" of Cavalieri's Principle, called the *Shell Method.*

To illustrate the idea see the figure below, where the x-th "radial slice" or "shell" is highlighted. The volume $dV(x)$ of this shell is clearly

$$dV(x) = 2\pi x h \, dx.$$

Just cut the shell vertically and stretch it out. The shell then becomes a thin rectangular slab (parallelepiped) with base $2\pi x$, height h and thickness dx, which has volume $2\pi x h \, dx$. The total volume of the cylinder therefore is the integral

$$\begin{aligned} \int_0^r dV &= 2\pi h \int_0^r x \, dx \\ &= \left(\pi h x^2\right)\Big|_0^r = \pi r^2 h. \end{aligned}$$

More general, let $c \in \mathbb{R}$, $c \leq a < b$, and let f and g be continuous on $[a, b]$ such that $g(x) \leq f(x)$ for all $x \in [a, b]$. Let Ω denote the solid generated by revolving the region between f and g over $[a, b]$ about the axis $x = d$. Then the volume differential of Ω at $x \in [a, b]$ is given by

$$dV(x) = 2\pi(x - d)\big(f(x) - g(x)\big)dx,$$

and the total volume by

$$\begin{aligned} V &= \int_a^b dV \\ &= 2\pi \int_a^b (x - d)\big(f(x) - g(x)\big)dx, \end{aligned}$$

Example 4.8
Find the volume of the solid generated by revolving the region R between $y = x^2$ and $y = 2x$ located in the I-th quadrant about the y-axis.

SOLUTION: Using the "radial version" of Theorem 4.21, it suffices to compute the area $A(x)$ of the x-th "radial" cross-section of the given solid of revolution for $x \in [0,2]$. As Figure 4.22 shows, each such cross-section constitutes the surface of a cylinder of radius x and height $2x - x^2$. Its area is therefore

$$A(x) = 2\pi x(2x - x^2) = 2\pi(2x^2 - x^3),$$

while the volume of the x-th "radial slice" or "shell" of the solid is

$$dV(x) = 2\pi(2x^2 - x^3)\,dx.$$

The total volume of the solid is the integral

$$V = 2\pi \int_0^2 dV(x) = 2\pi \int_0^2 \left(2x^2 - x^3\right) dx$$
$$= 2\pi\left(\tfrac{2}{3}x^3 - \tfrac{1}{4}x^4\right)\Big|_0^2 = 2\pi\left(\tfrac{16}{3} - 4\right) = \tfrac{8}{3}\pi.$$

◁

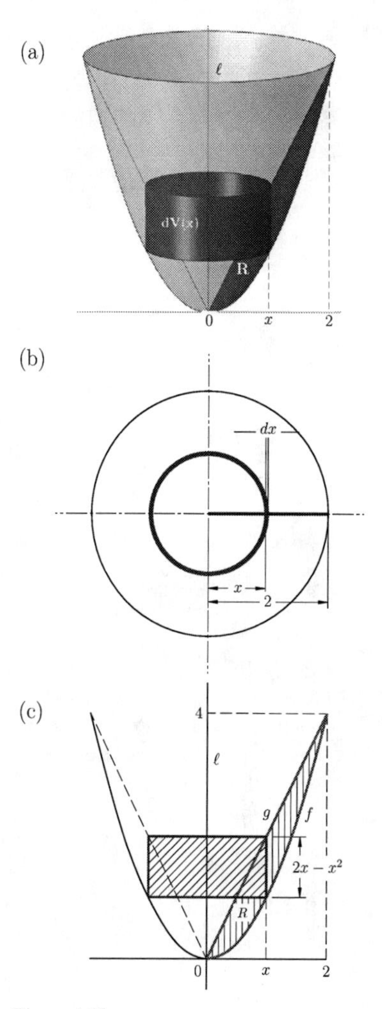

Figure 4.13:

Example 4.9
A torus of major radius a and minor radius b is obtained by revovling the region enclosed by the circle $(x-a)^2 + y^2 = b^2$ about the y-axis. Find the volume of the torus using the shell-method.

SOLUTION: Using the "radial version" of Cavalieri's Principle, we first compute the area $A(x)$ of the x-th radial cross-section of the given solid for $x \in [a-b, a+b]$. The circumference of this cross-section is $2\pi x$ while it height is $2y = 2\sqrt{b^2 - (x-a)^2}$, hence

$$A(x) = 4\pi x\sqrt{b^2 - (x-a)^2}$$

while the volume $dV(x)$ of the x-th radial slice is

$$dV(x) = 4\pi x\sqrt{b^2 - (x-a)^2}\,dx.$$

The total volume of the solid is therefore the integral

$$V = 4\pi \int_{a-b}^{a+b} x\sqrt{b^2 - (x-a)^2}\,dx$$

which with the substitution $z = x - a$ is equal to

$$= 4\pi \int_{-b}^{b} (z+a)\sqrt{b^2 - z^2}\,dz$$
$$= 4\pi \int_{-b}^{b} z\sqrt{b^2 - z^2}\,dz$$
$$+ 4\pi a \int_{-b}^{b} \sqrt{b^2 - z^2}\,dz$$

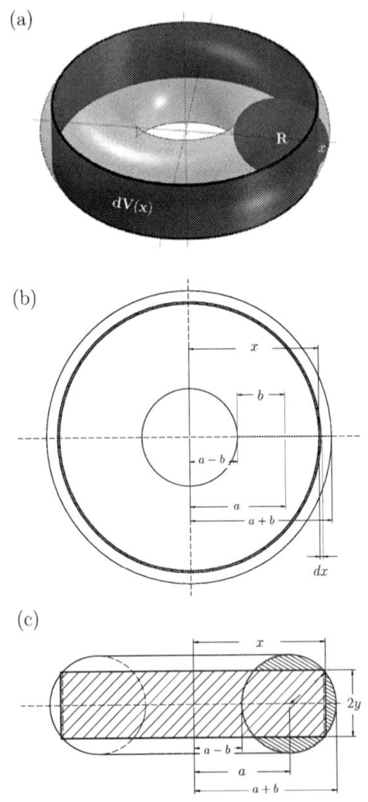

Figure 4.14:

which equals

$$= 4\pi a \cdot \tfrac{1}{2}\pi b^2 = 2\pi^2 ab^2,$$

since the integrand of the first integral is odd and the integral therefore equals zero and the second integral represents the area of a semi-circle of radius b.

$\triangleleft$

Arc Length

Definition 4.14 (Arc, Smooth Curve)
If f is continuously differentiable on $[a, b]$ (i.e. f' is continuous) then the curve

$$C := C_f := \{(x, f(x)) \mid a \le x \le b\}$$

is called an *arc* or a *smooth curve.*

Definition 4.15 (Rectifiable)
Let C_f be a curve, defined by the function f on $[a, b]$, and let $P := \{x_0, \ldots, s_n\}$ be a partition of $[a, b]$. Moreover, let

$$\Delta s_j := \sqrt{(x_j - x_{j-1})^2 + \big(f(x_j) - f(x_{j-1})\big)^2}$$

for $j = 1, \ldots, n$. Then,

$$s(P) := \sum_{j=1}^{n} \Delta s_j$$

gives the length of the polygonial arc inscribed in C_f relative to the partition P.
If the set $\{s(P) \mid P \text{ partition of } [a, b]\}$ is bounded above, then C_f is called *rectifiable* and we define the *length* s of C_f by

$$s := \sup\{s(P) \mid P \text{ partition of } [a, b]\}.$$

Remark 4.9 (Arc Length Differential)
If the function f defining the curve C_f is differentiable, then by the Mean Value Theorem

$$\Delta s_j = \sqrt{1 + \big(f'(\xi_j)\big)}$$

for some $\xi_j \in (x_{j-1}, x_j)$. Its supremum taken over all partitions of $[a, b]$

$$ds := \sqrt{1 + \big(f'(x)\big)}\, dx$$

is called the *arc length differential* of the curve C_f.

Theorem 4.22 (Arc Length)
If f is continuously differentiable, then the arc C_f is rectifiable and the total length of the curve C_f is given by

$$s = \int_a^b ds = \int_a^b \sqrt{1 + \left(f'(x)\right)^2}\, dx.$$

Area of a Surface of Revolution

Theorem 4.23
Let $c \in \mathbb{R}$ and let f be continuously differentiable on $[a, b]$ and $f(x) \geq c$ for all $x \in [a, b]$. Moreover, let Γ denote the surface generated by revolving the curve C_f about the axis $y = c$. Then the surface area differential of Γ at $x \in [a, b]$ is given by

$$\begin{aligned} dS &= 2\pi\big(f(x) - c\big)\, ds(x) \\ &= 2\pi\big(f(x) - c\big)\sqrt{1 + \left(f'(x)\right)^2}\, dx \end{aligned}$$

and the total surface area by

$$S = \int_a^b dS = \int_a^b 2\pi\big(f(x) - c\big)\sqrt{1 + \left(f'(x)\right)^2}\, dx.$$

Mass from Density

Definition 4.16 (Mass of a Rod)
Suppose the projection of the material solid Ω onto the real number line is given by the interval $[a, b]$, and suppose $\delta(x)$ is the density of the solid at $x \in [a, b]$ (units: mass per unit length). Then the mass differential of Ω at $x \in [a, b]$ is given by

$$dm(x) = \delta(x)\,dx,$$

and the total mass of Ω by

$$m = \int_a^b dm = \int_a^b \delta(x)\,dx.$$

Definition 4.17 (The Mass of a Plate)
Suppose the projection of the material solid Ω in the plane is the region between f and g over the interval $[a, b]$. Moreover, suppose that the density of the solid is given by the (1-dimensional) function $\delta : [a, b] \to \mathbb{R}$ (units: mass per unit area); note: we assume the density is constant along the lines $x = c$ for $c \in [a, b]$. Then the mass differential of Ω at $x \in [a, b]$ is given by

$$dm(x) = \big(f(x) - g(x)\big)\delta(x)\,dx,$$

and the total mass of Ω by

$$m = \int_a^b dm = \int_a^b \big(f(x) - g(x)\big)\delta(x)\,dx.$$

Definition 4.18 (The Mass of a Wire)
Suppose the axis of the wire W is a smooth curve C_f defined by the function f continuously differentiable on $[a, b]$. Moreover, suppose that $\delta(x)$ is the density of W at $x \in [a, b]$ (units: mass per unit length). Then the mass differential of W at $x \in [a, b]$ is given by

$$dm(x) = \delta(x)\,ds(x) = \delta(x)\sqrt{1 + \big(f'(x)\big)^2}\,dx,$$

and the total mass of W by

$$m = \int_a^b dm = \int_a^b \delta(x)\,ds = \int_a^b \delta(x)\sqrt{1 + \big(f'(x)\big)^2}\,dx.$$

Definition 4.19 (General $3d$-Case)
Suppose Ω is a material solid desribed by the 3-dimensional region B in $\mathbb{R}^3$ with density function δ assigning to each $(x, y, z) \in B$ its density $\delta(x, y, z)$. Then the mass differential of Ω at (x, y, z) is given by

$$dm(x, y, z) = \delta(x, y, z)\, dV(x, y, z)$$

and the the total mass of Ω by

$$m = \int_B dm = \int_B \delta(x, y, z)\, dV.$$

Center of Mass and Centroid

Definition 4.20 (The Center of Mass of a Rod)
Given a rod Ω with density function $\delta : [a, b] \to \mathbb{R}$. We consider the moment of Ω with respect to the point $x = 0$. The moment differential of Ω at $x \in [a, b]$ (the moment of the x-th slice of Ω) is given by

$$dM(x) = x\, dm(x) = x\delta(x)\, dx$$

and the total moment of Ω by

$$M = \int_a^b dM = \int_a^b x\, dm(x) = \int_a^b x\delta(x)\, dx.$$

The *center of mass* of Ω is the unique point $\overline{x} \in \mathbb{R}$ such that $\overline{x}m = M$ where m denotes the total mass of Ω. Thus

$$\overline{x} = \frac{\int dM}{\int dm} = \frac{\int_a^b x\delta(x)\, dx}{\int_a^b \delta(x)\, dx}.$$

Definition 4.21 (The Centroid of a Rod)
If the density of the rod Ω is constant, it is called a *uniform* or *homogeneous rod.* Its center of mass $\overline{x}$ is called the *centroid* of the rod. Clearly,

$$\overline{x} = \frac{1}{2}(a + b).$$

Definition 4.22 (The Center of Mass of a Plate)
Suppose the Plate Ω is given by the continuous functions f and g on $[a, b]$ and the density function δ on $[a, b]$. We consider the moment M_x about the x-axis and the moment M_y about the y-axis. Then the moment differentials of Ω at x with respect to the x-axis and y-axis are given by

$$dM_x(x) = \hat{y}\, dm(x) = \tfrac{1}{2}\big(f(x) + g(x)\big)\, dm(x),$$
$$dM_y(x) = \hat{x}\, dm(x) = x\, dm(x),$$

respectively, where $\hat{x}$ and $\hat{y}$ denote the x- and y-coordinate of the center of mass of the (uniform rod) mass differential $dm(x)$ at x. The total moments of Ω with respect to the x- and y-axis are given by

$$M_x = \int_a^b dM_x = \tfrac{1}{2}\int_a^b \Big(\big[f(x)\big]^2 - \big[g(x)\big]^2\Big)\delta(x)\, dx,$$
$$M_y = \int_a^b dM_y = \int_a^b x\big(f(x) - g(x)\big)\delta(x)\, dx,$$

respectively. The *center of mass* of Ω is the unique point $(\overline{x}, \overline{y}) \in \mathbb{R}^2$ such that $\overline{x}m = M_y$ and $\overline{y}m = M_x$, where m denotes the total mass of Ω. Thus

$$\overline{x} = \frac{\displaystyle\int dM_y}{\displaystyle\int dm} = \frac{\displaystyle\int \hat{x}\, dm}{\displaystyle\int dm}, \qquad \overline{y} = \frac{\displaystyle\int dM_x}{\displaystyle\int dm} = \frac{\displaystyle\int \hat{y}\, dm}{\displaystyle\int dm},$$

respectively.

Definition 4.23 (The Centroid of a Region)
If the density of the plate Ω is constant, we call its center of mass, the *centroid of the region* covered by the plate.

Definition 4.24 (The Center of Mass of a Wire)
Suppose the axis of the wire W is the smooth curve C_f, f continuously differentiable on $[a,b]$ with density δ on $[a,b]$. The mass differential $dm(x)$ has center of mass $(\hat{x},\hat{y}) = (x,y)$. Then the moment differential of W at x with respect to the x-axis and y-axis is given by

$$dM_x(x) = \hat{y}\,dm(x) = y\,dm(x) = y\,\delta(x)\,ds,$$
$$dM_y(x) = \hat{x}\,dm(x) = x\,dm(x) = x\,\delta(x)\,ds,$$

respectively, and the total moment of W with respect to the x-axis and y-axis by

$$M_x = \int_a^b dM_x = \int_a^b y\,\delta(x)\,ds(x).$$
$$M_y = \int_a^b dM_y = \int_a^b x\,\delta(x)\,ds(x),$$

respectively. The center of mass $(\overline{x},\overline{y})$ of W is therefore given by

$$\overline{x} = \frac{\int dM_y}{\int dm} = \frac{\int \hat{x}\,dm}{\int dm} = \frac{\int_a^b x\,\delta(x)\,ds}{\int_a^b \delta(x)\,ds},$$

$$\overline{y} = \frac{\int dM_x}{\int dm} = \frac{\int \hat{y}\,dm}{\int dm} = \frac{\int_a^b y\,\delta(x)\,ds}{\int_a^b \delta(x)\,ds},$$

Definition 4.25 (The Centroid of an Arc)
If the density of a wire is constant, it is called the *centroid of the arc* covered by the wire.

Definition 4.26 (Center of Mass in Three Dimensions)
Suppose Ω is a material solid desribed by the 3-dimensional region B in $\mathbb{R}^3$ with density function δ assigning to each $(x, y, z) \in B$ its density $\delta(x, y, z)$. Let $(\hat{x}, \hat{y}, \hat{z}) \in \mathbb{R}^3$ denote the center of mass of the mass differential dm at $(x, y, z) \in B$, then the center of mass $(\overline{x}, \overline{y}, \overline{z})$ of Ω is given by

$$\overline{x} = \frac{\displaystyle\int_B \hat{x}\, dm}{\displaystyle\int dm} = \frac{\displaystyle\int_B \hat{x}\, \delta(x, y, z)\, dV}{\displaystyle\int_B \delta(x, y, z)\, dV}$$

$$\overline{y} = \frac{\displaystyle\int_B \hat{y}\, dm}{\displaystyle\int dm} = \frac{\displaystyle\int_B \hat{y}\, \delta(x, y, z)\, dV}{\displaystyle\int_B \delta(x, y, z)\, dV}$$

$$\overline{z} = \frac{\displaystyle\int_B \hat{z}\, dm}{\displaystyle\int dm} = \frac{\displaystyle\int_B \hat{z}\, \delta(x, y, z)\, dV}{\displaystyle\int_B \delta(x, y, z)\, dV}$$

Definition 4.27 (Center of Mass of a Surface)
Let Γ denote the surface of B and let again $(\hat{x}, \hat{y}, \hat{z})$ denote the center of mass of the mass differential $dm = \delta(z, y, z)\, dS$ at $(x, y, z) \in \Gamma$. Then the center of mass $(\overline{x}, \overline{y}, \overline{z})$ of Γ is given by

$$\overline{x} = \frac{\displaystyle\int_S \hat{x}\, dm}{\displaystyle\int dm} = \frac{\displaystyle\int_S \hat{x}\, \delta(x, y, z)\, dS}{\displaystyle\int_S \delta(x, y, z)\, dS}$$

$$\overline{y} = \frac{\displaystyle\int_S \hat{y}\, dm}{\displaystyle\int dm} = \frac{\displaystyle\int_S \hat{y}\, \delta(x, y, z)\, dS}{\displaystyle\int_S \delta(x, y, z)\, dS}$$

$$\overline{z} = \frac{\displaystyle\int_S \hat{z}\, dm}{\displaystyle\int dm} = \frac{\displaystyle\int_S \hat{z}\, \delta(x, y, z)\, dS}{\displaystyle\int_S \delta(x, y, z)\, dS}$$

Definition 4.28 (Symmetry and Centroid)
If the (1-, 2-, 3-dimensional) region Ω has axis of symmetry ℓ then the centroid $\overline{z}$ of Ω lies on ℓ.

Fluid Pressure

Definition 4.29 (Hydrostatic Pressure, Force)

1. The pressure p exerted by a fluid with weight-density δ at depth h is given by
$$p = \delta h.$$
2. The force F exerted by a fluid on a (horizontal) surface of area A at depth h is
$$F = pA = \delta h A.$$

Definition 4.30 (Hydrostatic Force on a Vertical Plate)
Suppose a vertical plate is submerged or partially submerged in a liquid of weight density δ with the liquid level at $x = c$. Let the submerged portion of the plate extend from a to b on the x axis. Let $w(x)$ be the width of the plate at x for $x \in [a, b]$ and assume that w is continuous. Then the *hydrostatic force* F on the plate due to the liquid is given by

$$F = \int_a^b \delta(c - x)w(x)\,dx$$

Work

Definition 4.31 (Work under a Constant Force)
The work W performed by an object moved a distance d along a straight line by a constant force F is
$$W = Fd.$$

Definition 4.32 (Work)
Suppose an object moves from a to b under the influence of a force f, where f is a continuous function on $[a, b]$. Then the *work* W done by the force on the object is defined by

$$W = \int_a^b f(x)\,dx$$

Money Stream

Definition 4.33 (Money Stream)
A *money stream* is a continuous function representing the flow of money as a function of time (units: dollars per unit time).

Definition 4.34 (Future Value)

1. Suppose an amount P is deposited into an account at r percent interest compounded continuously. Then its value B after T years is

$$B = Pe^{rT},$$

called the *future value* of P after T years at rate r.

2. Now suppose a money stream $P(t)$ is deposited into an account at r percent interest compounded continuously. Then the "future value differential" $dB(t)$ at time t is

$$dB(t) = P(t)\,dt\,e^{r(T-t)}$$

and the future value of the entire money stream $P(t)$ at time T is

$$B = \int_0^T dB(t) = \int_0^T P(t)e^{r(T-t)}\,dt.$$

4.1.9 The Theorems of Pappus

Theorem 4.24 (The First Theorem of Pappus)
Suppose the solid Ω is generated by revolving the region R about the line ℓ that does not intersect R. Let $\overline{r}$ denote the distance of the centroid of R from ℓ. Then

$$V = 2\pi\overline{r}A,$$

where A denotes the area of R.

Theorem 4.25 (The Second Theorem of Pappus)
Suppose the surface Γ is generated by revolving the arc C about the line ℓ that does not intersect C. Let $\overline{r}$ denote the distance of the centroid of C from ℓ. Then

$$S = 2\pi\overline{r}s,$$

where s denotes the length of C.

4.2 Exercises

4.2.1 Evaluating Integrals

Integration by Substitution

1. $\int x \cos x^2 \, dx$ ha

2. $\int \cos^{-4} x \sin x \, dx$ ha

3. $\int (x-1)\sqrt{x+1} \, dx$ ha

4. $\int_1^3 4x\sqrt{6-2x} \, dx$ ha

5. $\int_{-1}^2 \frac{x^2}{\sqrt{x+2}} \, dx$ ha

6. $\int \sqrt{1+\sqrt{x}} \, dx$ ha

7. $\int t^5\sqrt{1+t^2} \, dt$ a

8. $\int (x^2+3)(x^3+9x+1)^{\frac{1}{3}} \, dx$ a

9. $\int \frac{e^{\sqrt{x}}}{\sqrt{x}} \, dx$ a

10. $\int \frac{\sin t}{1+\cos t} \, dt$ a

11. $\int \frac{1+\sqrt{x+1}}{\sqrt{x+1}} \, dx$ a

12. $\int \frac{\tan(\ln t)}{t} \, dt$ a

13. $\int \frac{\sin^{-1} 2x}{\sqrt{1-4x^2}} \, dx$ has

14. $\int \frac{x}{\sqrt{x^2-4}} \, dx$ ha

15. $\int \frac{x}{\sqrt{4-x^2}} \, dx$ ha

16. $\displaystyle\int_0^{5/4} \frac{1}{\sqrt{25-4x^2}}\,dx$ ha

17. $\displaystyle\int \frac{1}{\left(1-2x^2\right)^{5/2}}\,dx$ ha

18. $\displaystyle\int \frac{x}{\sqrt{2x^2+12x+19}}\,dx$ ha

19. $\displaystyle\int \frac{e^x}{\sqrt{9-e^{2x}}}\,dx$ has

20. $\displaystyle\int \frac{x+3}{\sqrt{x^2+4x+13}}\,dx$ has

21. $\displaystyle\int \frac{x\,dx}{\sqrt{4-x^2}}$ a

22. $\displaystyle\int \sqrt{6x-x^2-8}\,dx$ a

23. $\displaystyle\int \frac{x}{\sqrt{x-1}}\,dx$ a

24. $\displaystyle\int \frac{\tan^2 x}{x-\tan x}\,dx$ a

25. $\displaystyle\int \frac{dx}{x^2\sqrt{a^2+x^2}}$ has

26. $\displaystyle\int x^2 e^{-x^3}\,dx$ ha

Integration by parts:

1. $\displaystyle\int xe^x\,dx$ has

2. $\displaystyle\int x^2(e^x-1)\,dx$ a

3. $\displaystyle\int x^3 3^x\,dx$ a

4. $\displaystyle\int x\sin x\,dx$ has

5. $\displaystyle\int e^x \sin x\,dx$ has

6. $\displaystyle\int \ln x\,dx$ a

7. $\displaystyle\int x^r \ln x \, dx, \quad r \in \mathbb{R}$ ha

8. $\displaystyle\int x \cos \pi x \, dx$ a

9. $\displaystyle\int \cos(\ln x) \, dx$ a

10. $\displaystyle\int x \arctan(x^2) \, dx$ a

11. $\displaystyle\int \frac{1}{x} \sin^{-1}(\ln x) \, dx$ a

12. $\displaystyle\int x \arctan(x) \, dx$ a

13. $\displaystyle\int x \arcsin(x) \, dx$ a

14. $\displaystyle\int \frac{\ln(x+1)}{\sqrt{x+1}} \, dx$ a

15. $\displaystyle\int \frac{1}{x^2} \left(\frac{x}{x+1} \right)^{\frac{1}{3}} \, dx$ a

16. $\displaystyle\int \cos \sqrt{x} \, dx$ a

17. $\displaystyle\int x \ln(x+1) \, dx$ a

18. $\displaystyle\int \frac{\sin^{-1} 2x}{\sqrt{1-4x^2}} \, dx$ a

19. $\displaystyle\int x^2 \cosh 2x \, dx$ a

20. $\displaystyle\int x^2 (2x-1)^{-7} \, dx$ a

21. $\displaystyle\int x \tan^{-1}(x^2) \, dx$ a

22. $\displaystyle\int \sin^{-1} 2x \, dx$ a

23. $\displaystyle\int x^2 \tan^{-1}(x^2) \, dx$ a

24. $\displaystyle\int \frac{1}{x} \sin^{-1}(\ln x^2) \, dx$ a

25. $\int \sin\sqrt{x}\, dx$ a

Trigonometric integrals:

1. $\int \tan^3 x \sec^3 x\, dx$ has

2. $\int \tan^3 x \sec^4 x\, dx$ a

3. $\int \tan^4 x\, dx$ a

4. $\int \sin^4(3t)\cos(3t)\, dt$ a

5. $\int \sin^4 x \cos^3 x\, dx$ has

6. $\int \sec^3(6x)\, dx$ a

7. $\int \cos^6 x\, dx$ has

8. $\int \sin^4 x \cos^4 x\, dx$ has

9. $\int (\cos 3x - \sin x)^2\, dx$ a

10. $\int \tan^6 x\, dx$ a

11. $\int \tan x \sec^{\frac{3}{2}} x\, dx$ a

12. $\int \cos^3 x \cos^2 x\, dx$ a

13. $\int \csc^3\left(\frac{1}{2}x\right)\, dx$ a

14. $\int \sin 5x \cos 2x\, dx$ a

15. $\int \sec^5 x\, dx$ a

16. $\int \cot^2 x \csc^2 x\, dx$ a

17. $\int \sin 2x \sin 3x\, dx$ ha

Partial fractions:

1. $\displaystyle\int \frac{7}{(x-2)(x+5)}\,dx$ a

2. $\displaystyle\int \frac{x}{(x+1)(x+2)(x+3)}\,dx$ a

3. $\displaystyle\int \frac{dx}{(x-1)^3}$ a

4. $\displaystyle\int \frac{x^5}{(x-2)^2}\,dx$ a

5. $\displaystyle\int \frac{2x^2+3}{x^2(x-1)}\,dx$ a

6. $\displaystyle\int \frac{x^2+3}{x^2-3x+2}\,dx$ a

7. $\displaystyle\int \frac{-3x}{x^2-x-12}\,dx$ a

8. $\displaystyle\int \frac{x-3}{x^2+4x+1}\,dx$ a

9. $\displaystyle\int \frac{7+6x}{(6x+1)^2}\,dx$ a

10. $\displaystyle\int \frac{x^3+x^2+x+3}{(x^2+1)(x^2+3)}\,dx$ a

11. $\displaystyle\int \frac{x^2-3x-1}{x^3+x^2-2x}\,dx$ a

12. $\displaystyle\int \frac{x^4-x^3-x-1}{x^3-x^2}\,dx$ a

13. $\displaystyle\int e^{2x}\tan^2(e^{2x})\sec^2(e^{2x})\,dx$ a

14. $\displaystyle\int \frac{1-\cos x}{1+\sin x}\,dx$ a

15. $\displaystyle\int \frac{25x^2+21+2x}{(x+3)(5x^2+3)}\,dx$ a

Improper integrals:

1. $\displaystyle\int_0^\infty \frac{1}{(2+x)^\pi}\,dx$ ha

2. $\displaystyle\int_2^\infty \frac{1}{x(\ln x)^3}\,dx$ ha

3. $\displaystyle\int_0^1 \frac{3x^2-1}{\sqrt[3]{x^3-x}}\,dx$ ha

4. $\displaystyle\int_0^1 \frac{dx}{\sqrt{1-x^2}}$ a

5. $\displaystyle\int_e^\infty \frac{\ln x}{x}\,dx$ a

6. $\displaystyle\int_0^1 \frac{e^{\sqrt{x}}}{\sqrt{x}}\,dx$ a

7. $\displaystyle\int_0^3 \frac{1}{x^2-6x+5}\,dx$ a

8. $\displaystyle\int_1^\infty \frac{1}{x(x+1)}\,dx$ a

9. $\displaystyle\int_0^1 \arcsin x\,dx$ a

4.2.2 Applications of Integration

Area between two curves

1. $y = 2x$, $y = 4 - x$, and $y = -x/2$
2. Under $|x^2 - 1|$ over $[-3, 1]$
3. Between $y = \cos x$ and $y = \sin x$ over $[0, \pi/4]$

Volume by slicing

1. Under $y = \sqrt{1 - x^2}$, rotated about the line $y = -1$
2. Under $y = \sqrt{1 - x^2}$, rotated about the line $x = 1$
3. The base of the solid is the circular region $x^2 + y^2 \leq 1$. A cross-section perpendicular to the y-axis is a square.

4. The base of the solid is the semi-circular region under $y = \sqrt{1-x^2}$ over $[-1,1]$. A cross-section perpendicular to the y-axis is an equilateral triangle.

5. Two cylinders of radius r intersect at right angles, their axes meeting. Find the volume of the intersection. What is the shape of a cross-section parallel to the plane of the two axis?

Volume by shells:

1. $\{(x,y) \mid 0 \le x \le 2,\ 0 \le y \le x^3\}$, rotated about the line $x = 3$

2. $\{(x,y) \mid 3 \le x \le 5,\ \sqrt{x} \le y \le x^2\}$, rotated about the line $x = 1$

3. $\{(x,y) \mid 3 \le x \le 5,\ \sqrt{x} \le y \le x^2\}$, rotated about the line $x = 5$

4. $\{(x,y) \mid 3 \le x \le 5,\ \sqrt{x} \le y \le x^2\}$, rotated about the line $y = 1$

5. Find the volume generated by revolving an isosceles right triangle with leg a about a line perpendicular to the hypotenuse and through a vertex (not the right-angle vertex).

Arc length:

1. $C : y = x^2, 0 \le x \le 2$

2. $C : y = \ln(\cos x), 0 \le x \le \pi/4$

Surface area:

1. $C := \{(x,y) \mid x^2 + y^2 = 1,\ x \ge 0,\ y \ge 0\}$ revolved about the line $y = 1$

2. $C := \{(x,y) \mid x^2 + y^2 = 1,\ x \ge 0,\ y \ge 0\}$ revolved about the line $x = 2$

3. $C : y = \cos x, 0 \le x \le \pi/2$

4. A square with side s is revolved about a line parallel to a diagonal and a distance r from the nearest vertex.

Mass from density:

1. Find the mass of a rod of length ℓ whose density varies as the square of the distance from one end.

2. Find the mass of a semi-circular disc of radius a in which density varies as the square of the distance from the diameter.

3. Find the mass of the solid circular cylinder of radius r and length ℓ whose density varies as the distance from the axis of symmetry.

4. Find the mass of a circular disc of radius 4 in which density varies as the distance from the center.

Center of mass and centroid:

1. Find the center of mass of a rod of length 2 whose density varies as the distance from one end.

2. Find the center of mass of a plate covering the region below $y = x^2$ over $[0, 2]$ whose density var ies as the distance from the line $x = 3$.

3. Find the center of mass of a wire covering the unit circle whose density varies as the distance from the line $x = 2$.

4. Find the centroid of a semi-circular disc of radius a

5. Find the centroid of a solid cone.

6. Find the centroid of a solid parabolic cone obtained by revolving the region below $y = \sqrt{x}$ over $[0, a]$ about the x-axis.

7. Find the centroid of a hemispherical bowl of radius a.

8. Find the centroid of the arc $y = \ln x$, $1 \leq x \leq e$.

9. Use Pappus' Theorem to find the volume of the solid generated by revolving the region below $y = \sin x$ over $[0, \pi]$ about the line $x = 2\pi$.

Fluid pressure:

1. A water tank is a cylinder 6 feet in diameter, standing on end. If water fills the tank to a depth of 4 feet, how much force is exerted on the sides of the tank?

2. The face of a dam is an isosceles trapezoid of height 20 feet and bases 10 feet and 40 feet. It is situated vertically with the smaller base at the bottom. Find the force on the dam if the water level is 2 feet below the top of the dam.

Work:

1. Find the work done in moving on object from $x = 1$ to $x = 5$ if the force acting at point x is given by $F(x) = \sqrt{x^2 + 1}$.

2. A vertical cylindrical tank of radius 2 feet and height 6 feet is full of water. Find the work done in pumping out the water (a) to an outlet at the top of the tank; (b) to a level 5 feet above the top of the tank. (Assume that the water weights 62.5 pounds per cubic foot.)

3. A conical container (vertex down) of radius r feet and height h feet is full of a liquid weighing σ pounds per cubic foot. Find the work done in pumping out the top $\frac{1}{2}h$ feet of liquid: (a) to the op of the tank; (b) to a level k feet about the top of the tank.

4. A chain that weights 15 pounds per foot is hanging from the top of an 80-foot building to the ground. How much work is done in pulling the chain to the top of the building?

4.2.3 Parametric Curves:

Converting Parametric Curves

Express the parametric curve by an equation in x and y:

1. $x(t) = 3t - 1, y(t) = 5 - 2t$

2. $x(t) = 2t - 1, y(t) = 8t^3 - 5$

3. $x(t) = e^{2t}, y(t) = e^{2t} - 1$

4. $x(t) = e^{6t}, y(t) = e^{3t} - 3$

5. $x(t) = 2 \sin t, y(t) = \cos t$

Parametrizing Curves

Find a parametrization $x = (t), y = y(t), t \in [0, 1]$:

1. The line segment from $(3, 7)$ to $(8, 5)$.

2. The line segment from $(2, 6)$ to $6, 3)$.

3. The parabolic arc $x = 1 - y^2$ from $(0, -1)$ to $(0, 1)$.

4. The curve $y^2 = x^3$ from $(4, 8)$ to $(1, 1)$.

The Slope of Parametric Curves

Find the slope of the given curve at the given point and give an equation of the tangent line:

1. $x(t) = 2\cos t$, $y(t) = 3\sin t$, $t = \pi/3$

2. $x(t) = 1 + t$, $y(t) = 1 - t^2$, $t = 2$

Arc Length of Parametric Curves

Find the length of the parametric arc:

1. $x(t) = t$, $y(t) = t^2$, $t \in [0, 3]$

2. $x(t) = 2\sin t$, $y(t) = 2\cos t$, $t \in [0, \pi/2]$

4.2.4 Polar Coordinates:

Converting Equations into Polar Coordinates

Write the equation in polar coordinates:

1. $x = 2$

2. $x^2 + (y - 1)^2 = 4$

3. $y = mx$

Converting Equations into Cartesian Coordinates

Write the equation in cartesian coordinates:

1. $r\sin\theta = 4$

2. $\tan\theta = 2$

3. $r = 4\sin(\theta + \pi)$

Graphs of Polar Curves

Sketch the polar curves:

1. $\theta = -\frac{1}{4}\pi$

2. $r = -2\sin\theta$

3. $r = 1 - \cos\theta$

4. $r = \cos 2\theta$

5. $r = \sin 3\theta$

6. $r = -1 + 2\cos\theta$

Area in Polar Coordinates

Calculate the area enclosed by the polar curve:

1. $r = a\cos\theta$, $\theta \in [-\frac{1}{2}\pi, \frac{1}{2}\pi]$

2. $r = a(1 + \cos 3\theta)$, $\theta \in [-\frac{1}{3}\pi, \frac{1}{3}\pi]$

3. Outside $r = 2$, but inside $r = 4\sin\theta$

4. Inside $r = 4$, and between the lines $\theta = \frac{1}{2}\pi$ and $r = 2\sec\theta$

5. Inside the inner loop of $r = 1 - 2\sin\theta$

Slope of Polar Curves

Find the slope of the polar curve:

1. $r = a$, $[a, \pi/3]$

2. $r = \sin 2\theta$, $[\sqrt{3}/2, \pi/6]$

3. $r = 1+\cos\theta$; where are the tangent lines either horizontal or vertical?

Arc Length of Polar Curves

Find the length of the polar curve:

1. $r = 3\cos\theta$

2. $r = a(1 + \cos\theta)$

4.3 Hints to the Exercises

4.3.1 Evaluating Integrals

Integration by Substitution

1. Substitute $y = x^2$
2. Substitute $y = \cos x$
3. Substitute $y = x + 1$
4. Substitute $y = 6 - 2x$
5. Substitute $y = x + 2$
6. Try the substitution (Theorem 4.18) given by $y = 1 + \sqrt{x}$.
13. Try the substitution (Theorem 4.18) given by $y = \sin^{-1}(2x)$.
14. Substitute $y = x^2 - 4$
15. Substitute $y = 4 - x^2$
16. Substitute $x = \frac{5}{2} \sin\theta$
17. Substitute $x = 1/\sqrt{2} \sin\theta$
18. First complete the square!
19. Reduce the integrand by the substitution $y = e^x$ (Theorem 4.18).
20. Complete the square and use trigonometric substitution (Remark 4.2:1).

25. Use trigonometric substitution (Remark 4.2:1) and set $x = a \tan\theta$.
26. let $u = -x^3$

Integration by parts:

1. Use integration by parts (Theorem 4.19) and set $u = x$ and $v' = e^x$.
4. Use integration by parts (Theorem 4.19) and set $u = x$ and $v' = \sin x$
5. Integrate by parts (Theorem 4.19) twice such that the remaining integral is a multiple of the original integral. Combine those integrals on the left side of the equation.
7. Distinguish the cases $r = -1$ and $r \neq -1$!

Trigonometric integrals:

1. Use the Second Pythagorean identity (Theorem 2.3:2) and rewrite

$$\begin{aligned}\tan^3 x \sec^3 x &= \sec^2 x \tan^2 x(\sec x \tan x)\\ &= \sec^2(\sec^2 x - 1)(\sec x \tan x).\end{aligned}$$

   Then apply Theorem 4.15:2.

5. Use the first Pythagorean identity (Theorem 2.3:1) and rewrite

$$\cos^3 x = \cos^2 x \cos x = (1 - \sin^2 x) \cos x.$$

   Then apply (Theorem 4.15:2).

7. You can use the trigonometric identity $\cos^2 x = \frac{1}{2}(1 + \cos 2x)$ to reduce the power of cos in the integrand or apply the recursion formula Theorem 4.16:9.

8. Use the trigonometric identities

$$\begin{aligned}\sin x \cos x &= \tfrac{1}{2}\sin(2x)\\ \cos^2 x &= \tfrac{1}{2}(1 + \cos(2x))\\ \sin^2 x &= \tfrac{1}{2}(1 - \cos(2x))\end{aligned}$$

   to reduce the number of factors.

17. Use Theorem 2.7:1

Partial fractions:

Improper integrals:

1. Substitute $y = 2 + x$

2. Substitute $y = \ln x$

3. Use Substitution

4.3.2 Applications of Integration

Area between two curves

Volume by slicing

Volume by shells:

Arc length:

Surface area:

Mass from density:

Center of mass and centroid:

Fluid pressure:

Work:

4.3.3 Parametric Curves:

Converting Parametric Curves

Express the parametric curve by an equation in x and y:

Parametrizing Curves

Find a parametrization $x = (t)$, $y = y(t)$, $t \in [0, 1]$:

The Slope of Parametric Curves

Find the slope of the given curve at the given point and give an equation of the tangent line:

Arc Length of Parametric Curves

Find the length of the parametric arc:

4.3.4 Polar Coordinates:

Converting Equations into Polar Coordinates

Write the equation in polar coordinates:

Converting Equations into Cartesian Coordinates

Write the equation in cartesian coordinates:

Graphs of Polar Curves

Sketch the polar curves:

Area in Polar Coordinates

Calculate the area enclosed by the polar curve:

Slope of Polar Curves

Find the slope of the polar curve:

Arc Length of Polar Curves

Find the length of the polar curve:

4.4 Answers to the Exercises

4.4.1 Evaluating Integrals

Integration by Substitution

1. $\frac{1}{2}\sin x^2 + C$
2. $\frac{1}{3}\cos^{-3} x + C$
3. $\frac{2}{5}(x+1)^{5/2} - \frac{4}{3}(x+1)^{3/2} + C$
4. $\frac{96}{5}$
5. $\frac{26}{15}$
6. $\frac{8}{15} + \frac{4}{15}(1+\sqrt{x})^{\frac{3}{2}}(3\sqrt{x-2}) + C$
7. $\frac{1}{7}t^4(1+t^2)^{\frac{3}{2}} - \frac{4}{35}t^2(1+t^2)^{\frac{3}{2}} + \frac{8}{105}(1+t^2)^{\frac{3}{2}} + C$
8. $\frac{1}{4}(x^3+9x+1)^{\frac{4}{3}} + C$
9. $2e^{\sqrt{x}} + C$
10. $-\ln|1+\cos(t)| + C$
11. $x + 2\sqrt{x+1} + C$
12. $-\ln|\cos(\ln(t))| + C$
13. $\frac{1}{4}\arcsin^2(2x) + C$
14. $\sqrt{x^2-4} + C$
15. $-\sqrt{4-x^2} + C$
16. $\frac{\pi}{12}$
17. $\dfrac{2x^3}{\left(1-2x^2\right)^{3/2}} + \dfrac{x}{\left(1-2x^2\right)^{1/2}} + C$
18. $\frac{1}{2}\sqrt{2x^2+12x+19} - \frac{1}{2}\sqrt{2}\ln\left|\sqrt{2x^2+12x+19} + \sqrt{2}(x+3)\right| + C$
19. $\sin^{-1}\left(\frac{1}{3}e^x\right) + C$
20. $\sqrt{x^2+4x+13} + \ln\left|x+2+\sqrt{x^2+4x+13}\right| + C$
21. $-\sqrt{4-x^2} + C$
22. $-\frac{1}{4}(6-2x)\sqrt{6x-x^2-8} + \frac{1}{2}\arcsin(x-3) + C$
23. $\frac{2}{3}(x-1)^{\frac{3}{2}} + 2\sqrt{x-1} + C$

24. $-\ln|x-\tan(x)|+C$

25. $-\dfrac{\sqrt{a^2+x^2}}{a^2x}+C$

26. $-\frac{1}{3}e^{-x^3}+c$

Integration by parts:

1. $xe^x - e^x + C$
2. $x^2e^x - 2xe^x + 2e^x + C$
3. $\frac{x^3\cdot 3^3}{\ln 3} - \frac{3x^2\cdot 3^x}{(\ln 3)^2} + \frac{6x\cdot 3^x}{(\ln 3)^3} - \frac{6\cdot 3^x}{(ln 3)^4} + C$
4. $\sin(x) - x\cos(x) + C$
5. $\frac{e^x}{2}\Big(\sin x - \cos x\Big) + C$
6. $x\ln(x) - x + C$
7. $\begin{cases} \frac{1}{2}\ln^2 x + C & r=-1 \\ x^{r+1}\left(\dfrac{\ln x}{r+1} - \dfrac{1}{(r+1)^2}\right) + C & r \neq -1 \end{cases}$
8. $\frac{1}{\pi^2}(\cos(\pi x) + \pi x\sin(\pi x) + C)$
9. $\frac{1}{2}\cos(\ln(x)) + \frac{1}{2}\sin(\ln(x)) + C$
10. $\frac{1}{2}x^2\arctan(x^2) - \frac{1}{4}\ln(x^4+1) + C$
11. $\ln(x)\arcsin(\ln(x)) + \sqrt{1-(\ln(x))^2} + C$
12. $\frac{1}{2}x^2\arctan(x) - \frac{1}{2}x + \frac{1}{2}\arctan(x) + C$
13. $\frac{1}{2}x^2\arcsin(x) + \frac{1}{4}x\sqrt{1-x^2} - \frac{1}{4}\arcsin(x) + C$
14. $2\ln|x+1|\sqrt{x+1} - 4\sqrt{x+1} + C$
15. $-\dfrac{3(x+1)}{2\sqrt[3]{x^2(x+1)}} + C$
16. $2\cos(\sqrt{x}) + 2\sqrt{x}\sin(\sqrt{x}) + C$
17. $\frac{1}{2}(x+1)^2\ln(x+1) - (x+1)\ln(x+1) - \frac{1}{4}x^2 + \frac{1}{2}x + \frac{3}{4} + C$
18. $\frac{1}{4}\arcsin^2(2x) + C$
19. $\frac{1}{2}x^2\sinh(2x) - \frac{1}{2}x\cosh(2x) + \frac{1}{4}\sinh(2x) + C$
20. $-\dfrac{1}{32(2x-1)^4} - \dfrac{1}{20(2x-1)^5} - \dfrac{1}{48(2x-1)^6} + C$

21. $\frac{1}{2}x^2\arctan(x^2) - \frac{1}{4}\ln(x^2+1) + C$

22. $x\arcsin(2x) + \frac{1}{2}\sqrt{1-4x^2} + C$

23. $\frac{1}{3}x^3\arctan(x) - \frac{1}{6}x^2 + \frac{1}{6}\ln(x^2+1) + C$

24. $\frac{1}{2}\ln(x^2)\arcsin(\ln(x^2)) + \frac{1}{2}\sqrt{1-(\ln(x^2))^2} + C$

25. $2\sin(\sqrt{x}) - 2\sqrt{x}\cos(\sqrt{x}) + C$

Trigonometric integrals:

1. $\frac{1}{5}\sec^5 x - \frac{1}{3}\sec^3 x + C$

2. $\frac{1}{6}\tan^6 x + \frac{1}{4}\tan^2 x + C$

3. $\frac{1}{3}\tan^3 x - \tan x + x + C$

4. $\frac{1}{5}\sin^5(3t) + C$

5. $\frac{1}{5}\sin^5 x - \frac{1}{7}\sin^7 x + C$

6. $\frac{1}{12}\sec 6x\tan 6x + \frac{1}{12}\ln|\sec 6x + \tan 6x| + C$

7. $\frac{5}{16}x + \frac{15}{64}\sin 2x + \frac{3}{64}\sin 4x + \frac{1}{192}\sin 6x + C.$

8. $\frac{3}{128}x - \frac{1}{128}\sin(4x) + \frac{1}{1024}\sin(8x)$

9. $\frac{1}{12}\sin(6x) + x + \frac{1}{4}\cos(4x) - \frac{1}{2}\cos(2x) - \frac{1}{4}\sin(2x) + C$

10. $\frac{1}{5}\tan^5 x - \frac{1}{3}\tan^3 x + \tan x - x + C$

11. $\frac{2}{3}\sec^{\frac{3}{2}}(x) + C$

12. $\frac{1}{2}\sin(x) + \frac{1}{40}\sin(5x) + \frac{1}{8}\sin(3x) + C$

13. $-\frac{\cos(\frac{1}{2}x)}{\sin^2(\frac{1}{2}x)} + \ln|\csc(\frac{1}{2}x) - (\cot(\frac{1}{2}x))| + C$

14. $-\frac{1}{4}\cos(7x) - \frac{1}{6}\cos(3x) + C$

15. $\frac{\sin(x)}{4\cos^4(x)} + \frac{3\sin(x)}{8\cos^2(x)} + \frac{3}{8}\ln|\sec(x) + \tan(x)| + C$

16. $-\frac{\cos^3(x)}{3\sin^3(x)} + C$

17. $-\frac{1}{10}\sin 5x + \frac{1}{2}\sin x + C$

Partial fractions:

1. $\ln|x-2| - \ln|x+5| + C$
2. $-\frac{3}{2}\ln|x+3| - \frac{1}{2}\ln|x+1| + 2\ln|x+2| + C$
3. $-\dfrac{1}{2(x-1)^2} + C$
4. $\frac{1}{4}x^4 + \frac{4}{3}x^3 + 6x^2 + 32x - \frac{32}{x-2} + 80\ln|x-2| + C$
5. $-3\ln|x| + 5\ln|x-1| + \frac{3}{x} + C$
6. $x + 7\ln|x-2| - 4\ln|x-1| + C$
7. $-\frac{9}{7}\ln|x+3| - \frac{12}{7}\ln|x-4| + C$
8. $\frac{1}{2}\ln|x^2+4x+1| + \frac{5\sqrt{3}}{3}\text{arctanh}(\frac{\sqrt{3}}{3}x + \frac{2\sqrt{3}}{3}) + C$
9. $-\dfrac{1}{6x+1} + \frac{1}{6}\ln|6x+1| + C$
10. $\arctan(x) + \frac{1}{2}\ln|x^2+3| + C$
11. $\frac{1}{2}\ln|x| - \ln|x-1| + \frac{3}{2}\ln|x+2| + C$
12. $\frac{1}{2}x^2 + 2\ln|x| - 2\ln|x-1| - \frac{1}{x} + C$
13. $\dfrac{\tan^3(e^{2x})}{6} + C$
14. $\ln|\tan^2(\frac{1}{2}x) + 1| - 2\ln|\tan(\frac{1}{2}x) + 1| - \frac{2}{\tan(\frac{1}{2}x)+1} + C$
15. $5\ln|x+3| + \frac{1}{10}\ln|5x^2+3| + \frac{1}{5}\arctan|\frac{x}{\sqrt{\frac{3}{5}}}| + C$

Improper integrals:

1. $\dfrac{2^{1-\pi}}{\pi-1}$
2. $\dfrac{1}{2(\ln 2)^2}$
3. 0
4. $\pi/2$
5. Diverges
6. $2e-2$
7. Diverges
8. $\ln 2$
9. $\pi/2 - 1$

4.4.2 Applications of Integration

Area between two curves

Volume by slicing

Volume by shells:

Arc length:

Surface area:

Mass from density:

Center of mass and centroid:

Fluid pressure:

Work:

4.4.3 Parametric Curves:

Converting Parametric Curves

Express the parametric curve by an equation in x and y:

Parametrizing Curves

Find a parametrization $x = (t)$, $y = y(t)$, $t \in [0, 1]$:

The Slope of Parametric Curves

Find the slope of the given curve at the given point and give an equation of the tangent line:

Arc Length of Parametric Curves

Find the length of the parametric arc:

4.4.4 Polar Coordinates:

Converting Equations into Polar Coordinates

Write the equation in polar coordinates:

Converting Equations into Cartesian Coordinates

Write the equation in cartesian coordinates:

Graphs of Polar Curves

Sketch the polar curves:

Area in Polar Coordinates

Calculate the area enclosed by the polar curve:

Slope of Polar Curves

Find the slope of the polar curve:

Arc Length of Polar Curves

Find the length of the polar curve:

4.5 Solutions to Selected Exercises

4.5.1 Evaluating Integrals

Integration by Substitution

13. Using the substitution (Theorem 4.18)

$$\begin{aligned} y &= \sin^{-1}(2x) \\ dy &= \frac{2dx}{\sqrt{1-4x^2}} \end{aligned}$$

and thus

$$\frac{dy}{2} = \frac{dx}{\sqrt{1-4x^2}}$$

follows immediately that

$$\begin{aligned} \int \frac{\sin^{-1} 2x}{\sqrt{1-4x^2}}\, dx &= \tfrac{1}{2} \int y\, dy \\ &= \tfrac{1}{4} y^2 + C \\ &= \tfrac{1}{4} \Big(\sin^{-1}(2x)\Big)^2 + C \\ &= \tfrac{1}{4} \sin^{-2}(2x) + C. \end{aligned}$$

19. The substitution (Theorem 4.18) $y = e^x$, $dy = e^x dx$ leads to the reduction

$$\int \frac{e^x}{\sqrt{9-e^{2x}}}\, dx = \int \frac{dy}{\sqrt{9-y^2}} =$$

This integral can be solved using trigonometric substitution (Remark 4.2:1). With the substitution $y = 3\sin\theta$, $d\theta = 3\cos\theta$ we obtain

$$\begin{aligned} &= \int \frac{3\cos\theta\, d\theta}{\sqrt{9 - 9\sin^2\theta}} = 3\int \frac{\cos\theta\, d\theta}{\sqrt{9(1-\sin^2\theta)}} \\ &= 3\int \frac{\cos\theta\, d\theta}{\sqrt{9\cos^2\theta}} = \int \frac{\cos\theta\, d\theta}{\cos\theta} \\ &= \int d\theta = \theta + C \end{aligned}$$

provided that $\cos\theta > 0$. Thus, solving the equation $e^x = y = 3\sin\theta$ for theta, we find that $\theta = \sin^{-1}\left(\frac{1}{3}e^x\right)$ and conclude

$$\int \frac{e^x}{\sqrt{9-e^{2x}}}\,dx = \sin^{-1}\left(\tfrac{1}{3}e^x\right) + C.$$

20. We first "complete the square" in the radicant: $x^2 + 4x + 13 = (x + 2)^2 + 9$. Then the integral

$$\int \frac{x+3}{\sqrt{x^2+4x+13}}\,dx = \\ = \int \frac{x+3}{\sqrt{(x+2)^2+9}}\,dx$$

and with the substitution (Theorem 4.18) $y = x + 2$, $dy = dx$

$$= \int \frac{y+1}{\sqrt{y^2+9}}\,dy$$

With the trigonometric substitution (Remark 4.2:1) $y = 3\tan\theta$ which implies $dy = 3\sec^2\theta\,d\theta$ the last integral changes to

$$\begin{aligned} &= \int \frac{3\tan\theta+1}{\sqrt{9\tan^2\theta+9}}(3\sec^2\theta)\,d\theta \\ &= 3\int \frac{3\tan\theta+1}{\sqrt{9(\tan^2\theta+1)}}(\sec^2\theta)\,d\theta \\ &= 3\int \frac{3\tan\theta+1}{\sqrt{9\sec^2\theta)}}(\sec^2\theta)\,d\theta \\ &= \int \frac{3\tan\theta+1}{\sec\theta}(\sec^2\theta)\,d\theta \end{aligned}$$

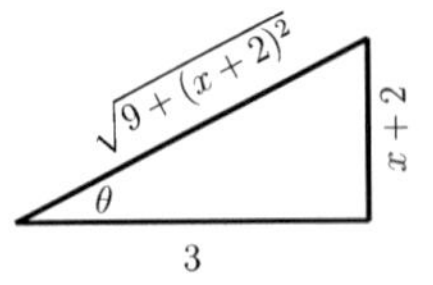

Figure 4.15:

provided $\sec\theta > 0$, recall that $\sqrt{A^2} = |A|$, and thus

$$\begin{aligned}
&= \int (3\tan\theta + 1)\sec\theta\, d\theta \\
&= 3\int \sec\theta\tan\theta\, d\theta + \int \sec\theta\, d\theta \\
&= 3\sec\theta + \ln|\sec\theta + \tan\theta| + C \\
&= \sqrt{x^2+4x+13} \\
&\quad + \ln\left|\tfrac{1}{3}(x+2+\sqrt{x^2+4x+13})\right| + C \\
&= \sqrt{x^2+4x+13} + \ln\left|\tfrac{1}{3}\right| \\
&\quad + \ln\left|x+2+\sqrt{x^2+4x+13}\right| + C
\end{aligned}$$

and then combining $\ln\left|\frac{1}{3}\right|$ and C to a new constant D, we finally obtain

$$\begin{aligned}
&= \sqrt{x^2+4x+13} \\
&\quad + \ln\left|x+2+\sqrt{x^2+4x+13}\right| + D
\end{aligned}$$

since $3\tan\theta = y = x+2$, implies that

$$\begin{aligned}
\tan\theta &= \tfrac{1}{3}(x+2) \\
\sec\theta &= \tfrac{1}{3}\sqrt{9+(x+2)^2} \\
&= \tfrac{1}{3}\sqrt{x^2+4x+13}
\end{aligned}$$

as can be easily verified consulting the following right triangle:

25. With the substitution $x = a\tan\theta$ and $dx = a\sec^2\theta\, d\theta$, cf. (Remark 4.2:1), follows

$$\begin{aligned}
&\int \frac{dx}{x^2\sqrt{a^2+x^2}}\, dx = \\
&= \int \frac{a\sec^2\theta\, d\theta}{a^2\tan^2\theta\sqrt{a^2+a^2\tan^2\theta}} \\
&= \int \frac{a\sec^2\theta\, d\theta}{a^2\tan^2\theta\sqrt{a^2(1+\tan^2\theta)}} \\
&= \int \frac{a\sec^2\theta\, d\theta}{a^2\tan^2\theta\sqrt{a^2\sec^2\theta)}} \\
&= \int \frac{a\sec^2\theta\, d\theta}{a^3\tan^2\theta\sec\theta} = \frac{1}{a^2}\int \frac{\sec\theta\, d\theta}{\tan^2\theta}
\end{aligned}$$

provided $a > 0$ and $\sec\theta > 0$ and thus

$$\begin{aligned}
&= \frac{1}{a^2}\int \frac{\frac{1}{\cos\theta}\,d\theta}{\frac{\sin^2\theta}{\cos^2\theta}} = \frac{1}{a^2}\int \frac{\cos\theta}{\sin^2\theta}\,d\theta \\
&= \frac{1}{a^2}\int \cot\theta\csc\theta\,d\theta \\
&= -\frac{1}{a^2}\csc\theta + C
\end{aligned}$$

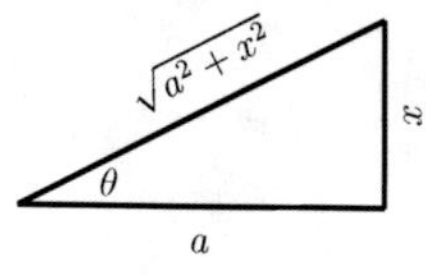

Figure 4.16:

by Theorem 4.16:8 and equals

$$= -\frac{\sqrt{a^2+x^2}}{a^2 x} + C$$

since $x = a\tan\theta$ implies $\tan\theta = \frac{x}{a}$ and thus

$$\csc\theta = \frac{\sqrt{a^2+x^2}}{x}$$

as can be easily verified consulting the triangle:

Integration by parts:

1. We use integration by parts (Theorem 4.19) in the form

$$\int uv' = uv - \int u'v$$

and try to reduce the factor x in the integrand to 1 by choosing $u = x$:

$$\begin{aligned}
\int xe^x\,dx &= \int \underbrace{x}_{u}\underbrace{e^x}_{v'}\,dx \\
&= \underbrace{x}_{u}\underbrace{e^x}_{v} - \int \underbrace{1}_{u'}\underbrace{e^x}_{v}\,dx \\
&= xe^x - \int e^x\,dx \\
&= xe^x - e^x + C
\end{aligned}$$

4. We use integration by parts (Theorem 4.19) in the form

$$\int uv' = uv - \int u'v$$

and try to reduce the factor x in the integrand to 1 by choosing $u = x$:

$$\begin{aligned}\int x \sin x \, dx &= \int \underbrace{x}_{u} \underbrace{\sin x}_{v'} \, dx \\ &= x(-\cos x) - \int 1 \cdot (-\cos x) \, dx \\ &= -x \cos x + \int \cos x \, dx\end{aligned}$$

which by Theorem 4.15:3) equals

$$= -x \cos x + \sin x + C.$$

5. We will use integration (Theorem 4.19) by parts twice and will see that the remaining integral will be a multiple of the original integral:

$$\begin{aligned}\int e^x \sin x \, dx &= \int \underbrace{e^x}_{u'} \underbrace{\sin x}_{v} \, dx \\ &= e^x \sin x - \int \underbrace{e^x}_{u'} \underbrace{\cos x}_{v} \, dx \\ &= e^x \sin x - \Big(e^x \cos x + \int e^x \sin x \, dx\Big) \\ &= e^x \big(\sin x - \cos x\big) - \int e^x \sin x \, dx\end{aligned}$$

Collecting the term $\int e^x \sin x \, dx$ on the left side, we obtain

$$2 \int e^x \sin x \, dx = e^x \big(\sin x - \cos x\big)$$

and thus

$$\int e^x \sin x \, dx = \tfrac{e^x}{2} \big(\sin x - \cos x\big) + C.$$

Trigonometric integrals:

1. Using the Second Pythagorean identity (Theorem 2.3:2), we obtain the equation

$$\begin{aligned}\tan^3 x \sec^3 x &= \sec^2 x \tan^2 x (\sec x \tan x) \\ &= \sec^2 (\sec^2 x - 1)(\sec x \tan x).\end{aligned}$$

Thus

$$\begin{aligned}\int \tan^3 x \sec^3 x \, dx &= \\ &= \int \sec^2(\sec^2 x - 1)(\sec x \tan x)\, dx \\ &= \int \sec^4 x(\sec x \tan x)\, dx \\ &\qquad - \int \sec^2 x(\sec x \tan x)\, dx\end{aligned}$$

which by the Generalized Powerrule (Theorem 4.15:2) equals

$$= \tfrac{1}{5}\sec^5 x - \tfrac{1}{3}\sec^3 x + C.$$

5. Using the first Pythagorean identity (Theorem 2.3:1), we obtain

$$\cos^3 x = \cos^2 x \cos x = (1 - \sin^2 x)\cos x.$$

Hence

$$\begin{aligned}\int \sin^4 x \cos^3 x \, dx &= \\ &= \int \sin^4 x(1 - \sin^2 x)\cos x \, dx \\ &= \int \sin^4 x \cos x \, dx - \int \sin^6 x \cos x \, dx\end{aligned}$$

which by the Generalized Powerrule (Theorem 4.15:2) equals

$$= \tfrac{1}{5}\sin^5 x - \tfrac{1}{7}\sin^7 x + C$$

7. *First Method:* We demonstrate first, how the integral can be solved by reducing the power of cos: To this end, we first use Theorem 2.6:2 and Theorem 2.7:3 and reduce $\cos^3 A$:

$$\begin{aligned}\cos^3 A &= \cos A \cos^2 A \\ &= \tfrac{1}{2}\cos A(1 + \cos 2A) \\ &= \tfrac{1}{2}\cos A + \tfrac{1}{2}\cos A \cos 2A \\ &\underset{(Th\ 2.7:3)}{=} \tfrac{1}{2}\cos A + \tfrac{1}{4}\cos A + \tfrac{1}{4}\cos 3A \\ &= \tfrac{3}{4}\cos A + \tfrac{1}{4}\cos 3A.\end{aligned}$$

Then

$$\cos^6 x = (\cos^2 x)^3 \underset{(Th\ 2.6:2)}{=} \left[\tfrac{1}{2}(1+\cos 2x)\right]^3$$
$$= \tfrac{1}{8}\left[1 + 3\cos 2x + 3\cos^2 2x + \cos^3 2x\right]$$

which by Theorem 2.6:2 and the computation above equals

$$= \tfrac{1}{8}\left[1 + 3\cos 2x + \tfrac{3}{2} + \tfrac{3}{2}\cos 4x \right.$$
$$\left. + \tfrac{3}{4}\cos 2x + \tfrac{1}{4}\cos 6x\right]$$
$$= \tfrac{1}{8}\left[\tfrac{5}{2} + \tfrac{15}{4}\cos 2x + \tfrac{3}{2}\cos 4x + \tfrac{1}{4}\cos 6x\right].$$

Therefore,

$$\int \cos^6 x\, dx = \tfrac{1}{8}\left[\tfrac{5}{2}x + \tfrac{15}{8}\sin 2x \right.$$
$$\left. + \tfrac{3}{8}\sin 4x + \tfrac{1}{24}\sin 6x\right] + C$$
$$= \tfrac{5}{16}x + \tfrac{15}{64}\sin 2x$$
$$+ \tfrac{3}{64}\sin 4x + \tfrac{1}{192}\sin 6x + C.$$

Second Method: We will now show how the integral can be solved using the recurrence formula Theorem 4.16:7

$$\int \cos^n x\, dx = \tfrac{1}{n}\cos^{n-1} x \sin x$$
$$+ \tfrac{n-1}{n}\int \cos^{n-2} x\, dx.$$

Note that the recurrence allows us to compute $\int \cos^n x\, dx$ provided we have already computed $\int \cos^{n-2} x\,dx$.

$$\int \cos^0 x\,dx = x$$
$$\int \cos^1 x\,dx = \sin x$$
$$\begin{aligned}
\int \cos^2 x\,dx &= \tfrac{1}{2}\cos x \sin x + \tfrac{1}{2}\int \cos^0 x\,dx \\
&= \tfrac{1}{2}\cos x \sin x + \tfrac{1}{2}x
\end{aligned}$$
$$\begin{aligned}
\int \cos^3 x\,dx &= \tfrac{1}{3}\cos^2 x \sin x + \tfrac{2}{3}\int \cos^1 x\,dx \\
&= \tfrac{1}{3}\cos^2 x \sin x + \tfrac{2}{3}\sin x
\end{aligned}$$
$$\begin{aligned}
\int \cos^4 x\,dx &= \tfrac{1}{4}\cos^3 x \sin x + \tfrac{3}{4}\int \cos^2 x\,dx \\
&= \tfrac{1}{4}\cos^3 x \sin x \\
&\qquad + \tfrac{3}{4}\big(\tfrac{1}{2}\cos x \sin x + \tfrac{1}{2}x\big) \\
&= \tfrac{1}{4}\cos^3 x \sin x + \tfrac{3}{8}\cos x \sin x + \tfrac{3}{8}x
\end{aligned}$$
$$\begin{aligned}
\int \cos^5 x\,dx &= \tfrac{1}{5}\cos^4 x \sin x \\
&\qquad + \tfrac{4}{5}\big(\tfrac{1}{3}\cos^3 x \sin x + \tfrac{2}{3}\sin x\big) \\
&= \tfrac{1}{5}\cos^4 x \sin x \\
&\qquad + \tfrac{4}{15}\cos^2 x \sin x + \tfrac{8}{15}\sin x
\end{aligned}$$
$$\begin{aligned}
\int \cos^6 x\,dx &= \tfrac{1}{6}\cos^5 x \sin x \\
&+ \tfrac{5}{6}\big(\tfrac{1}{4}\cos^3 x \sin x + \tfrac{3}{8}\cos x \sin x + \tfrac{3}{8}\big) \\
&= \tfrac{1}{6}\cos^5 x \sin x \\
&+ \tfrac{5}{24}\cos^3 x \sin x + \tfrac{5}{16}\cos x \sin x + \tfrac{5}{16}x.
\end{aligned}$$

8. We will use the trigonometric identities

$$\sin x \cos x = \tfrac{1}{2}\sin(2x) \tag{4.1}$$
$$\cos^2 x = \tfrac{1}{2}\big(1 + \cos(2x)\big) \tag{4.2}$$
$$\sin^2 x = \tfrac{1}{2}\big(1 - \cos(2x)\big) \tag{4.3}$$

to reduce the number of factors of the integrand (at the expense of creating higher "frequencies").

$$\begin{aligned}
&\int \sin^4 x \cos^4 x \, dx = \\
&\underset{(4.1)}{=} \int \left(\tfrac{1}{2}\sin 2x\right)^4 = \tfrac{1}{16}\int \left(\sin^2 2x\right)^2 dx \\
&\underset{(4.3)}{=} \tfrac{1}{16}\int \left(\tfrac{1}{2}(1-\cos 4x)\right)^2 dx \\
&= \tfrac{1}{64}\int (1 - 2\cos 4x + \cos^2 4x) dx \\
&\underset{(4.2)}{=} \tfrac{1}{64}\int (1 - 2\cos 4x + \tfrac{1}{2} + \tfrac{1}{2}\cos 8x) dx \\
&= \tfrac{1}{64}\int (\tfrac{3}{2} - 2\cos 4x + \tfrac{1}{2}\cos 8x) dx \\
&= \tfrac{1}{64}(\tfrac{3}{2}x - \tfrac{1}{2}\sin 4x + \tfrac{1}{16}\sin 8x) + C \\
&= \tfrac{3}{128}x - \tfrac{1}{128}\sin 4x + \tfrac{1}{1024}\sin 8x + C.
\end{aligned}$$

Partial fractions:

Improper integrals:

4.5.2 Applications of Integration

Area between two curves

Volume by slicing

Volume by shells:

Arc length:

Surface area:

Mass from density:

Center of mass and centroid:

Fluid pressure:

Work:

4.5.3 Parametric Curves:

Converting Parametric Curves

Express the parametric curve by an equation in x and y:

Parametrizing Curves

Find a parametrization $x = (t)$, $y = y(t)$, $t \in [0, 1]$:

The Slope of Parametric Curves

Find the slope of the given curve at the given point and give an equation of the tangent line:

Arc Length of Parametric Curves

Find the length of the parametric arc:

4.5.4 Polar Coordinates:

Converting Equations into Polar Coordinates

Write the equation in polar coordinates:

Converting Equations into Cartesian Coordinates

Write the equation in cartesian coordinates:

Graphs of Polar Curves

Sketch the polar curves:

Area in Polar Coordinates

Calculate the area enclosed by the polar curve:

Slope of Polar Curves

Find the slope of the polar curve:

Arc Length of Polar Curves

Find the length of the polar curve:

5

SEQUENCES AND SERIES

5.1 Definitions and Results

5.1.1 Sequences of Real Numbers

Definition 5.1 (Real Sequence)
A real-valued function $a : \mathbb{N} \to \mathbb{R}$ from the set of positive integers $\mathbb{N}$ into the real numbers $\mathbb{R}$ is called a *sequence of real numbers* or a *real sequence.* It is customary to denote $a(n)$ simply by a_n and the whole sequence by $(a_n)_{n\in\mathbb{N}} = (a_1, a_2, a_3, \ldots)$ or by $\{a_n\}_{n\in\mathbb{N}} = \{a_1, a_2, a_3, \ldots\}$. Often the notation is simplified to (a_n) or $\{a_n\}$.

Definition 5.2 (Boundedness)
A (real) sequence $(a_n)_{n\in\mathbb{N}}$ is said to be

1. *bounded from above* if there exists some constant $B \in \mathbb{R}$ such that $a_n \leq B$ for all $n \in \mathbb{N}$.;
2. *bounded from below* if there exists some constant $b \in \mathbb{R}$ such that $a_n \geq b$ for all $n \in \mathbb{N}$;
3. *bounded* if there exists some constant $C \in \mathbb{R}$ such that $|a_n| \leq C$ for all $n \in \mathbb{N}$.

Definition 5.3 (Monotonicity)
A (real) sequence $(a_n)_{n\in\mathbb{N}}$ is said to be

1. *increasing* if there exists some $k \in \mathbb{N}$ such $a_n < a_{n+1}$ for all $n \in \mathbb{N}$ with $n > k$.
2. *non-decreasing* if there exists some $k \in \mathbb{N}$ such $a_n \leq a_{n+1}$ for all $n \in \mathbb{N}$ with $n > k$.
3. *decreasing* if there exists some $k \in \mathbb{N}$ such $a_n > a_{n+1}$ for all $n \in \mathbb{N}$ with $n > k$.
4. *non-increasing* if there exists some $k \in \mathbb{N}$ such $a_n \geq a_{n+1}$ for all $n \in \mathbb{N}$ with $n > k$.

Definition 5.4 (Limit)
A (real) sequence $(a_n)_{n\in\mathbb{N}}$ has *limit* $L \in \mathbb{R}$, written

$$\lim_{n\to\infty} a_n = L, \quad a_n \to L,$$

if and only if, for any $\varepsilon > 0$ there exists some constant $N \in \mathbb{N}$ such that for all $n \in \mathbb{N}$:

$$n > N \Rightarrow |a_n - L| < \varepsilon.$$

We then also say that the sequence $(a_n)_{n\in\mathbb{N}}$ *converges* to L, or is *convergent*. If no such L exists, the sequence is said to *diverge* or be *divergent*.

Theorem 5.1 (Limit Theorem)
Suppose $(a_n)_{n\in\mathbb{N}}$ and $(b_n)_{n\in\mathbb{N}}$ are convergent sequences and $\alpha \in \mathbb{R}$, then

1. $\lim_{n\to\infty} (a_n + b_n) = \lim_{n\to\infty} a_n + \lim_{n\to\infty} b_n$
2. $\lim_{n\to\infty} (\alpha a_n) = \alpha \lim_{n\to\infty} a_n$
3. $\lim_{n\to\infty} (a_n b_n) = \big(\lim_{n\to\infty} a_n\big)\big(\lim_{n\to\infty} b_n\big)$
4. $\lim_{n\to\infty} \frac{a_n}{b_n} = \frac{\lim_{n\to\infty} a_n}{\lim_{n\to\infty} b_n}$, *if $b_n \neq 0$ for all $n \in \mathbb{N}$ and $\lim_{n\to\infty} b_n \neq 0$.*

Theorem 5.2 (Continuity)
Suppose the function $f : D_f \to \mathbb{R}$ is continuous and $(a_n)_{n\in\mathbb{N}}$ a convergent sequence such that $a_n \in D_f$ for all $n \in \mathbb{N}$. Then the sequence of images $\big(f(a_n)\big)_{n\in\mathbb{N}}$ also converges and

$$\lim_{n\to\infty} f(a_n) = f\Big(\lim_{n\to\infty} a_n\Big).$$

Theorem 5.3 (Squeezing Theorem)
Suppose $a_n \leq b_n \leq c_n$ for all $n \in \mathbb{N}$ with $n > k$ for some $k \in \mathbb{N}$. If $\lim_{n\to\infty} a_n = \lim_{n\to\infty} c_n = L$, then $\lim_{n\to\infty} b_n = L$.

Example 5.1
If $\lim_{n\to\infty} a_n = 0$ and $(b_n)_{n\in\mathbb{N}}$ is bounded, then

$$\lim_{n\to\infty} a_n b_n = 0.$$

Theorem 5.4 (Convergence ⇒ Boundedness)
Every convergent sequence is bounded.

Theorem 5.5 (Monotonicity + Boundedness ⇒ Convergence)
Every bounded monotonic sequence is convergent.
More precisely: Every non-decreasing sequence that is bounded above converges to its least upper bound, every non-increasing sequence that is bounded below converges to its greatest lower bound.

Example 5.2
If the sequence $(a_n)_{n\in\mathbb{N}}$ is monotonic and unbounded, then

$$\lim_{n\to\infty} \frac{1}{a_n} = 0.$$

Theorem 5.6 (Some Important Limits)

1. $\lim_{n\to\infty} c = c, \ c \in \mathbb{R};$
2. $\lim_{n\to\infty} x^{1/n} = 1, \ x > 0;$
3. Geometric Sequence:
$$\lim_{n\to\infty} x^n = \begin{cases} 0 & |x| < 1 \\ 1 & x = 1 \\ \textit{diverges} & \textit{else}; \end{cases}$$
4. $\lim_{n\to\infty} \frac{1}{n^p} = 0, \ p > 0;$
5. $\lim_{n\to\infty} \frac{x^n}{n!} = 0;$
6. $\lim_{n\to\infty} \frac{\ln n}{n} = 0;$
7. $\lim_{n\to\infty} n^{1/n} = 1;$
8. $\lim_{n\to\infty} \left(1 + \frac{x}{n}\right)^n = e^x;$

5.1.2 Infinite Series

Definition 5.5 (Infinite Series)
Let $(a_n)_{n\in\mathbb{N}}$ be a (real) sequence. The *infinite series* $\sum_{k=1}^{\infty} a_k$ is defined to be the sequence $(S_n)_{n\in\mathbb{N}}$ of *n-th partial sums*

$$S_n := \sum_{k=1}^{n} a_k.$$

The coefficients a_n are called the *terms* of the series.
Moreover, the series $\sum_{k=1}^{\infty} a_k$ is called a *series of positive terms* if $a_k > 0$ for all $k \in \mathbb{N}$.

Definition 5.6 (Convergence)
Let $\sum_{k=1}^{\infty} a_k$ be an infinite series and $S \in \mathbb{R}$. The series $\sum_{k=1}^{\infty} a_k$ *converges to* S if the sequence $(S_n)_{n\in\mathbb{N}}$ of partial sums converges to S. If $\lim_{n\to\infty} S_n = S$, we call S the *value* of the infinite series and write

$$\sum_{k=1}^{\infty} a_k = S.$$

Definition 5.7 (Absolute Convergence)
The series $\sum_{k=1}^{\infty} a_k$ is said to be *absolutely convergent* if the series $\sum_{k=1}^{\infty} |a_k|$ converges.

Definition 5.8 (Conditional Convergence)
If the series $\sum_{k=1}^{\infty} a_k$ converges, but not absolutely, then it is called *conditionally convergent.*

Theorem 5.7 (Combinations of Series)
If $\sum_{k=1}^{\infty} a_k$ and $\sum_{k=1}^{\infty} b_k$ are convergent series and $\alpha \in \mathbb{R}$, then

1. $\displaystyle\sum_{k=1}^{\infty}(a_k + b_k) = \sum_{k=1}^{\infty} a_k + \sum_{k=1}^{\infty} b_k;$

2. $\displaystyle\sum_{k=1}^{\infty} \alpha a_k = \alpha \sum_{k=1}^{\infty} a_k.$

5.1.3 Convergence Tests

Theorem 5.8 (Divergence Test)
If the series $\sum_{n=1}^{\infty} a_n$ converges, then $\lim_{n\to\infty} a_n = 0$. Equivalently: If $\lim_{n\to\infty} a_n \neq 0$ then $\sum_{n=1}^{\infty} a_n$ diverges.

Theorem 5.9 (Integral Test)
If f is a positive, continuous and decreasing function on the interval $[K, \infty)$ for some constant $K \in \mathbb{N}$, then

$$\sum_{k=K}^{\infty} f(k) \text{ conv.} \Leftrightarrow \int_K^{\infty} f(x)\,dx \text{ conv.}$$

Corollary 5.1 (p-Series Test)
The p-series

$$\sum_{n=K}^{\infty} \frac{1}{n}$$

converges if and only if $p > 1$, for any $K \in \mathbb{N}$.

Theorem 5.10 (Comparison Test)
Suppose the series of positive terms $\sum_{k=1}^{\infty} b_k$ dominates the series of positive terms $\sum_{k=1}^{\infty} a_k$, i.e. $b_k \geq a_k$ for all $k > K$ for some $K \in \mathbb{N}$. Then

1. *If $\sum_{k=1}^{\infty} b_k$ converges, then $\sum_{k=1}^{\infty} a_k$ converges;*
2. *If $\sum_{k=1}^{\infty} a_k$ diverges, then $\sum_{k=1}^{\infty} b_k$ diverges;*

Theorem 5.11 (Limit Comparison Test)
Suppose $\sum_{k=1}^{\infty} a_k$ and $\sum_{k=1}^{\infty} b_k$ are series of positive terms and

$$0 < \lim_{n\to\infty} \frac{a_n}{b_n} < \infty$$

then

$$\sum_{k=1}^{\infty} a_k \text{ conv.} \Leftrightarrow \sum_{k=1}^{\infty} b_k \text{ conv.}$$

Definition 5.9 (Alternating Series)
If $(a_n)_{n\in\mathbb{N}}$ is a sequence and $a_n \geq 0$ then the infinite series

$$\sum_{n=1}^{\infty} (-1)^{n+1} a_n$$

is called an *alternating series*.

Theorem 5.12 (Alternating Series Test, Leibniz Criterion)
If $(a_n)_{n\in\mathbb{N}}$ is decreasing, positive and $\lim_{n\to\infty} a_n = 0$, then

1. $\sum_{n=1}^{\infty}(-1)^{n+1}a_n$ *converges;*

2. $|S - S_n| < a_{n+1}$ *for all $n \in \mathbb{N}$, where S denotes the value of the alternating series.*

Theorem 5.13 (Ratio Test)
Let $\sum_{n=1}^{\infty} a_n$ be a series and let

$$\varrho := \lim_{n\to\infty}\left|\frac{a_{n+1}}{a_n}\right|$$

then

1. *If $\varrho < 1$ then $\sum_{n=1}^{\infty} a_n$ converges absolutely;*

2. *If $\varrho > 1$ then $\sum_{n=1}^{\infty} a_n$ diverges;*

3. *If $\varrho = 1$ then the test is inconclusive.*

Theorem 5.14 (Root Test)
Let $\sum_{n=1}^{\infty} a_n$ be a series and let

$$\varrho := \lim_{n\to\infty} |a_n|^{1/n}$$

then

1. *If $\varrho < 1$ then $\sum_{n=1}^{\infty} a_n$ converges absolutely;*

2. *If $\varrho > 1$ then $\sum_{n=1}^{\infty} a_n$ diverges;*

3. *If $\varrho = 1$ then the test is inconclusive.*

Theorem 5.15 (Important Series, Estimate)

1. Geometric Series:

$$\sum_{n=0}^{\infty} r^n = \begin{cases} \dfrac{1}{1-r} & |r| < 1 \\ \textit{diverges} & \textit{else} \end{cases}$$

2. p-series:

$$\sum_{n=1}^{\infty} \frac{1}{n^p} \text{ converges} \Leftrightarrow p > 1$$

3. Telescoping Series:

$$\sum_{n=1}^{\infty} \frac{1}{n(n+1)} = 1$$

4. Useful Estimate: *For all $k \in \mathbb{N}$ there exists a constant b_k such that for $x \in (b_k, \infty)$*

$$\ln x \le x^{1/k}$$

5.1.4 Power Series

Definition 5.10 (Power Series)

1. An infinite series of the form

$$\sum_{n=0}^{\infty} b_n x^n$$

is called a *power series* in x.

2. If g is any function, then an infinite series of the form

$$\sum_{n=0}^{\infty} b_n \big(g(x)\big)^n$$

is called a *power series* in $g(x)$.

Theorem 5.16 (Radius of Convergence, Interval of Convergence)
Let $\sum_{n=0}^{\infty} b_n x^n$ be a power series in x. Then there exists a unique r with $0 \leq r \leq \infty$ (note: r can be ∞ !) such that

1. $\displaystyle\sum_{n=0}^{\infty} b_n x^n$ *converges absolutely for all* $|x| < r$;

2. $\displaystyle\sum_{n=0}^{\infty} b_n x^n$ *diverges for all* $|x| > r$. *This r is called the* radius of convergence *of* $\sum_{n=0}^{\infty} b_n x^n$. *The interval*

$$\{x \in \mathbb{R} \,\Big|\, \sum_{n=0}^{\infty} b_n x^n \text{ converges}\}$$

is called the interval of convergence *of the power series* $\sum_{n=0}^{\infty} b_n x^n$.

Theorem 5.17 (Computing the Radius of Convergence)
If $\sum_{n=0}^{\infty} b_n x^n$ is a power series in x, then its radius of convergence is given by

1. $r = \displaystyle\lim_{n\to\infty} \left| \frac{b_n}{b_{n+1}} \right|$

2. $r = \dfrac{1}{\displaystyle\lim_{n\to\infty} \sqrt[n]{|b_n|}}$ *(Hadamard)*

Theorem 5.18 (Differentiation and Integration)
Suppose $\sum_{n=0}^{\infty} b_n x^n$ is a power series with radius of convergence $r > 0$. Let $f(x) = \sum_{n=0}^{\infty} b_n x^n$. Then

1. *f is differentiable on $(-r, r)$ and its derivative f' has a power series representation with the same radius of convergence. Moreover,*

$$f'(x) = \sum_{n=1}^{\infty} n b_n x^{n-1}$$

for all $x \in (-r, r)$.

2. *f is integrable on $(-r, r)$ and its integral $\int f(x)\, dx$ has a power series representation with the same radius of convergence. Moreover,*

$$\int f(x)\, dx = \sum_{n=0}^{\infty} \frac{b_n}{n+1} x^{n+1} + C$$

for all $x \in (-r, r)$.

Corollary 5.2 (Taylor Coefficients)
f has continuous derivatives of all orders on $(-r, r)$ and

$$b_n = \frac{f^{(n)}(0)}{n!}$$

for all $n \in \{0, 1, 2, \ldots\}$.

Theorem 5.19 (Identity Theorem)
Suppose

$$\sum_{n=0}^{\infty} a_n x^n = \sum_{n=0}^{\infty} b_n x^n$$

for all x in the intersection of the intervals of convergence of the two series. Then $a_n = b_n$ for all $n \in \{0, 1, 2, \ldots, \}$.

Theorem 5.20 (Algebraic Operations)
Given two power series $f(x) = \sum_{k=0}^{\infty} a_k x^k$ and $g(x) = \sum_{k=0}^{\infty} b_k x^k$ with positive radius of convergence. Let $r > 0$ be chosen such that both power series converge for all $|x| < r$. Then

1. $f(x) + g(x) = \sum_{k=0}^{\infty} (a_k + b_k) x^k.$

2. $(\alpha f)(x) = \sum_{k=0}^{\infty} (\alpha a_k) x^k.$

3. $f(x) \cdot g(x) = \sum_{k=0}^{\infty} \left(\sum_{j=0}^{k} a_j b_{k-j} \right) x^k.$

Figure 5.1: *Brook Taylor (1685–1731)*

5.1.5 Taylor Series

Definition 5.11 (Taylor Series)
Suppose f possessess derivatives of any order on the interval I containing x_0. Then the series

$$T(x) := \sum_{j=0}^{\infty} \frac{f^{(j)}(x_0)}{j!} (x - x_0)^j$$

is called the *Taylor series* or *Taylor expansion* of f centered at x_0;

Remark 5.1
Note that the Taylor series T of a function that is infinitely differentiable, say

at $x_0 = 0$, does not have to converge f for some open interval I containing 0. The classic counterexample is

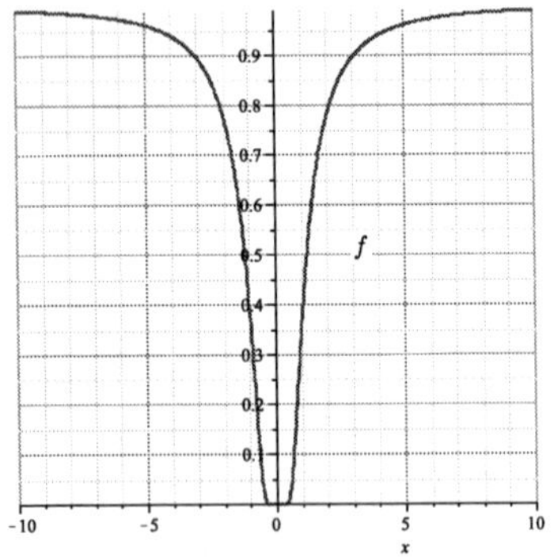

Figure 5.2: *Counterexample*

$$f(x) := \begin{cases} e^{-1/x^2} & \text{for } x \neq 0 \\ 0 & \text{else} \end{cases}$$

It can be shown that

$$f^{(j)}(0) = 0 \quad \text{for all } n = 0, 1, 2, \dots$$

Hence, $T(x) = 0$ for $x \in \mathbb{R}$, which implies that the Taylor series T of f converges to f only at $x_0 = 0$. Taylors remainder term provides a useful to tool to check whether the Taylor series of a function represents the function.

Definition 5.12 (Taylor Polynomials)
The n-th partial sum

$$T_n(x) = \sum_{j=0}^{n} \frac{f^{(j)}(x_0)}{j!}(x - x_0)^j$$

of the Taylor series expansion of f at x_0 is called the n-th *Taylor polynomial* of f centered at x_0 (which exists if f has derivatives up to order n on I).

Definition 5.13 (Taylor's Remainder Term)
Suppose P_n is the n-th Taylor polynomial of f centered at x_0. Then the term

$$R_n(x) := f(x) - T_n(x)$$

for $x \in I$ is called the n-*Taylor's remainder* for f at x.

Theorem 5.21 (Taylor's Remainder Theorem)
If f possesses derivatives up to order $n+1$ on the open interval I containing x_0, then for each $x \in I$

$$R_n(x) = \frac{1}{n!} \int_{x_0}^{x} f^{(n+1)}(t)(x - t)^n \, dt.$$

Theorem 5.22 (Lagrange's Form of Taylor's Remainder)
If f possesses derivatives up to order $n+1$ on the open interval I containing x_0, then for each $x \in I$ there exists a number ξ between x and x_0 such that

$$R_n(x) = \frac{f^{(n+1)}(\xi)}{(n+1)!}(x - x_0)^{n+1}$$

Theorem 5.23
Let T denote the Taylor series of f. Then

$$T(x) = f(x) \quad \Leftrightarrow \quad \lim_{n\to\infty} R_n(x) = 0$$

where $R_n(x)$ denotes the n-Taylor remainder for f at x.

Theorem 5.24 (Important Taylor Series Expansions)

1. Exponential Series:
$$e^x = \sum_{k=0}^{\infty} \frac{x^k}{k!}, \quad |x| < \infty$$
2. Sine Series:
$$\sin x = \sum_{k=0}^{\infty} (-1)^k \frac{x^{2k+1}}{(2k+1)!}, \quad |x| < \infty$$
3. Cosine Series:
$$\cos x = \sum_{k=0}^{\infty} (-1)^k \frac{x^{2k}}{(2k)!}, \quad |x| < \infty$$
4. Hyperbolic Sine Series:
$$\sinh x = \sum_{k=0}^{\infty} \frac{x^{2k+1}}{(2k+1)!}, \quad |x| < \infty$$
5. Hyperbolic Cosine Series:
$$\cosh x = \sum_{k=0}^{\infty} \frac{x^{2k}}{(2k)!}, \quad |x| < \infty$$
6. Geometric Series:
$$\frac{1}{1-x} = \sum_{k=0}^{\infty} x^k, \quad |x| < 1$$
7. Binomial Series:
$$(1+x)^\alpha = \sum_{k=0}^{\infty} \binom{\alpha}{k} x^k, \quad |x| < 1, \alpha \in \mathbb{R}$$
8. Logarithmic Series:
$$\ln(x+1) = \sum_{k=1}^{\infty} (-1)^{k+1} \frac{x^k}{k}, \quad |x| < 1$$
9. Arc-Tangens Series:
$$\arctan(x) = \sum_{k=0}^{\infty} (-1)^k \frac{x^{2k+1}}{2k+1}, \quad |x| < 1$$

5.2 Exercises

5.2.1 Sequences

Boundedness and Monotonicity

Investigate whether the following sequences are bounded, monotonic (and convergent):

1. $\left(\frac{n^2-1}{n^2}\right)_n$ a
2. $\left((-1)^n \frac{2}{n}\right)_n$ a
3. $\left(\frac{n^2-1}{n-1}\right)_n$ a
4. $\left(\frac{\ln n}{n-1}\right)_n$ h s $\left(\frac{\ln n}{\sqrt{n}}\right)_n$ bounded by 0 from below and by 1 from above, decreasing (starting at about 10, converges to 0
5. $\left(\frac{2^n}{(n+1)!}\right)_n$ a

Convergence

Determine the following limits if they exist:

1. $\lim_{n\to\infty} (n+2)^{\frac{1}{n}}$ a
2. $\lim_{n\to\infty} \frac{5^{n+1}}{4^{2n-1}}$ a
3. $\lim_{n\to\infty} \left(-\frac{1}{3}\right)^n$ a
4. $\lim_{n\to\infty} \left(\frac{1}{n} - n\right)$ a
5. $\lim_{n\to\infty} 2^n$ a
6. $\lim_{n\to\infty} \ln \frac{1}{n}$ a
7. $\lim_{n\to\infty} \left(1 + \frac{1}{3n}\right)^n$ ha
8. $\lim_{n\to\infty} \frac{x^{50n}}{n!}$ a
9. $\lim_{n\to\infty} \frac{\ln n^2}{n}$ a

10. $\lim_{n\to\infty} \ln\left(\frac{n-1}{n}\right)$ a

11. $\lim_{n\to\infty} \frac{2n!}{3n}$ a

12. $\lim_{n\to\infty} \left(1+\frac{1}{n}\right)^{2n^2}$ a

13. $\lim_{n\to\infty} (\sqrt{n+2}-\sqrt{n})$ a

14. $\lim_{n\to\infty} (\sqrt{n^2+n-1}-n)$ a

15. $\lim_{n\to\infty} \frac{1+2+3+\cdots+n}{n^2}$ a

16. The Finacci sequence $(a_n)_n$ is defined recursively by $a_1 := 1$, $a_2 := 1$ and $a_{n+1} := a_n + a_{n-1}$ for $n = 2, 3, 4 \ldots$. Calculate the first 20 terms of the sequence $(a_{n+1}/a_n)_n$. Does this sequence converge? What is its limit? a

5.2.2 Infinite Series

Limit

Find the sum of the infinite series if it exists:

1. $\sum_{n=1}^{\infty} \frac{3}{n(n+3)}$ a

2. $\sum_{n=1}^{\infty} \frac{1}{2n(n+1)}$ a

3. $\sum_{n=0}^{\infty} \frac{1-2^n}{3^n}$ a

4. $\sum_{n=1}^{\infty} \frac{3^n}{2^{n+2}}$ a

5. $\sum_{n=1}^{\infty} \left(\frac{1}{e^n}+\frac{1}{n(n+1)}\right)$ a

6. $\sum_{n=0}^{\infty} (-1)^n \frac{5^n}{3^{2n+3}}$

7. $\sum_{n=0}^{\infty} \left(\frac{1}{n+1}-\frac{1}{n+5}\right)$

Repeated Decimals

Rewrite each repeating decimal as a fraction:

1. $0.2222\ldots$ a
2. $0.232\,232\,232\ldots$ a
3. $27.56123123123\ldots$ a
4. $1.5342424242\ldots$ a

Convergence

Determine whether the following series converge. Justify your conclusion!

1. $\displaystyle\sum_{n=3}^{\infty} \frac{1}{\sqrt[5]{n}}$ a
2. $\displaystyle\sum_{n=1}^{\infty} \frac{2}{3+\ln n}$ a
3. $\displaystyle\sum_{n=0}^{\infty} \frac{1}{(3n+1)^5}$ a
4. $\displaystyle\sum_{n=1}^{\infty} \frac{1}{1+2+3+\cdots+n}$ a
5. $\displaystyle\sum_{n=0}^{\infty} \left(1+\frac{1}{n}\right)^n$ a
6. $\displaystyle\sum_{n=1}^{\infty} \frac{n!}{n^n}$ a
7. $\displaystyle\sum \frac{3n}{n^2+5}$ a
8. $\displaystyle\sum \frac{3n}{n^3+5}$ a
9. $\displaystyle\sum \frac{2}{2^n-1}$ a
10. $\displaystyle\sum \frac{3}{n-3^n}$ a
11. $\displaystyle\sum \frac{1}{\sqrt{n}\ln n}$ a

12. $\sum \frac{\sqrt{n}}{\ln n}$

13. $\sum \frac{3}{n^2 \ln n}$

14. $\sum \frac{\ln n}{n}$

15. $\sum_{k=1}^{\infty} \frac{k!}{k^k}$

16. $\sum \frac{\ln n}{n\sqrt{n}}$

17. $\sum \frac{\ln n}{n^2\sqrt{n}}$

18. $\sum \frac{\ln n}{n^{1.005}}$

19. $\sum \frac{2n-4}{2n}$

20. $\sum \frac{\sin^2}{n\sqrt{n}}$

21. $\sum \frac{n-1}{n^2 2^n}$

22. $\sum \sin^2(\pi n)$

Absolute and Conditional Convergence

Determine whether each infinite series converges absolutely, conditionally or diverges:

1. $\sum_{n=1}^{\infty} (-1)^{n+1} \frac{1}{2n+1}$ a

2. $\sum_{n=1}^{\infty} (-1)^n \frac{n^n}{n!}$ a

3. $\sum_{n=1}^{\infty} (-1)^n \frac{n+2}{n^3+3n+5}$ a

4. $\sum_{n=1}^{\infty} \frac{\cos n\pi}{\sqrt{n}}$ a

5. $\sum_{n=2}^{\infty}(-1)^n \frac{\ln n}{n}$ a

6. $\sum_{n=1}^{\infty}(-1)^{n+1} \frac{10^n}{(n+1)4^{2n+1}}$ a

7. $\sum \frac{(-1)^n \ln n}{2n-3}$ a

8. $\sum_{n=2}^{\infty}(-1)^{n+1} \frac{1}{n(\ln n)^2}$ a

9. $\sum_{n=1}^{\infty}(-1)^n \frac{1 \times 3 \times 5 \times \cdots \times (2n+1)}{2 \times 5 \times 8 \times \cdots \times (3n+2)}$ a

5.2.3 Power Series

Interval of Convergence

Find the radius of convergence and the interval of convergence of the power series:

1. $\sum_{k=1}^{\infty} \frac{k-1}{k} x^k$
2. $\sum_{k=2}^{\infty} \frac{1}{\ln k} x^k$
3. $\sum_{n=1}^{\infty} n^3 x^n$
4. $\sum_{n=2}^{\infty} \frac{n(n-2)}{2} x^n$
5. $\sum_{k=2}^{\infty} \ln(k)\, x^k$
6. $\sum_{n=1}^{\infty} (-1)^n \frac{1}{n^3} x^{n+3}$
7. $\sum_{n=1}^{\infty} \frac{n(n+2)}{(n+1)(3n+1)} x^n$
8. $\sum_{n=2}^{\infty} \frac{1}{\ln n} (x-3)^n$

9. $\displaystyle\sum_{n=1}^{\infty} \frac{(-1)^n}{\ln(n+1)}(x-1)^n$

10. $\displaystyle\sum_{n=1}^{\infty} x^{2n+1}$

11. $\displaystyle\sum_{n=0}^{\infty} \frac{n^3+1}{2^n}x^{5n}$

12. $\displaystyle\sum_{n=0}^{\infty} \left(2-\sqrt{n+1}\right)(x+1)^{3n}$

13. $\displaystyle\sum_{n=0}^{\infty} \frac{9^n\sqrt{n+1}}{2+n}(x-1)^{2n+1}$

Radius of Convergence

Find the radius of convergence:

1. $\displaystyle\sum_{n-1}^{\infty} \frac{n^n}{n!}x^n$

2. $\displaystyle\sum_{n=1}^{\infty} \frac{n^{2n}}{(2n)!}x^n$

3. $\displaystyle\sum_{n=1}^{\infty} \frac{1\times 3\times 5\times\cdots\times(2n-1)}{2^n[1\times 4\times 7\times\cdots\times(3n-2)]}x^n$

Limit Function

Find the sum of each power series and give its interval of convergence:

1. $\displaystyle\sum_{n=0}^{\infty} n(n-1)(n-2)x^{n-3}$

2. $\displaystyle\sum_{n=0}^{\infty} (-1)^n n x^n$

3. $\displaystyle\sum_{n=0}^{\infty} (-1)^n n(n-1)x^n$

4. $\displaystyle\sum_{n=0}^{\infty} \frac{(-1)^n}{n+1}x^{n+1}$

5. $\displaystyle\sum_{n=0}^{\infty} \frac{(-1)^n}{(n+1)(n+2)}x^{n+2}$

Expansions

Establish the following indenties:

1. $\frac{\pi}{4} = 1 - \frac{1}{3} + \frac{1}{5} - \frac{1}{7} + \frac{1}{9} + \cdots$

2. $\ln 2 = \frac{1}{1 \cdot 2} + \frac{1}{2 \cdot 2^2} + \frac{1}{3 \cdot 2^3} + \frac{1}{4 \cdot 2^4} + \cdots$

3. $\int e^{x^2}\, dx = \sum_{n=0}^{\infty} \frac{x^{2n+1}}{(2n+1)n!} + C$

Limit Function

Determine the limit function:

1. $1 - \frac{2}{2} + \frac{3}{4} - \frac{4}{8} + \frac{5}{16} - \cdots$

2. $1 + \frac{1}{2 \cdot 2} + \frac{1}{3 \cdot 2^2} + \frac{1}{4 \cdot 2^3} + \frac{1}{5 \cdot 2^4} + \cdots$

3. $\frac{1 \cdot 2}{1} + \frac{2 \cdot 3}{3} + \frac{3 \cdot 4}{3^2} + \frac{4 \cdot 5}{3^3} + \frac{5 \cdot 6}{3^4} + \cdots$

Taylor Series Expansions

Find the Taylor expansion at x_0 for each function and give its interval of convergence:

1. $f(x) = \frac{1}{1 - x^4}, \quad x_0 = 0$

2. $f(x) = \frac{1}{1 + x^3}, \quad x_0 = 0$

3. $f(x) = \frac{1}{(1 + x)^3}, \quad x_0 = 0$

4. $f(x) = \ln(1 + x^2), \quad x_0 = 0$

5. $f(x) = \frac{1}{1 + 2x}, \quad x_0 = 0$

6. $f(x) = \tan^{-1} 3x^2, \quad x_0 = 0$

7. $f(x) = \sin(3x), \quad x_0 = 0$

8. $f(x) = x^3 \cos x, \quad x_0 = 0$

9. $f(x) = x^3 e^{-3x^3}, \quad x_0 = 0$

10. $f(x) = \dfrac{1}{x}, \quad x_0 = -1$

11. $f(x) = \dfrac{1}{(1+2x)^4}, \quad x_0 = 2$

12. $f(x) = \sqrt{x}, \quad x_0 = 1$

13. $f(x) = \ln x, \quad x_0 = 2$

5.3 Anwers to the Exercises

5.3.1 Sequences

Boundedness and Monotonicity

Investigate whether the following sequences are bounded, monotonic (and convergent):

1. bounded by 0 from below and 1 from above, increasing, converges to 1
2. bounded by -2 from below and 1 from above, not monotonic (alternating), converges to 0
3. bounded by 2 from below, unbounded from above, increasing, diverges (converges to ∞) $\left(\frac{\ln n}{\sqrt{n}}\right)_n$ bounded by 0 from below and by 1 from above, decreasing (starting at about 10, converges to 0
5. bounded from below by 0 and 1 from above, decreasing, converges to 0

Convergence

Determine the following limits if they exist:

1. 1
2. 0
3. 0
4. diverges
5. diverges (converge to ∞)
6. diverges (converges to $-\infty$)
7. $\sqrt[3]{e}$
8. 0
9. 0
10. 0
11. diverges (converges to ∞)
12. diverges (converges to ∞)
13. 0
14. $1/2$
15. $1/2$

16. In rounded decimals: 1.5, 1.66667, 1.6000, 1.62500, 1.61538, 1.61905, 1.61765, 1.61818, 1.61798, 1.61806, 1.61803, 1.61804, 1.618303, 1,61803, 1,61803, 1,61803, 1,61803, 1,61803 ...; the sequences converges to $\frac{1}{2}(1+\sqrt{5}) \approx 1.6180339887498948482\ldots$

5.3.2 Infinite Series

Limit

Find the sum of the infinite series if it exists:

1. $11/6$
2. $1/2$
3. $-3/2$
4. diverges
5. $\dfrac{e}{e-1}$

Repeated Decimals

Rewrite each repeating decimal as a fraction:

1. $2/9$
2. $\dfrac{232}{999}$
3. $\dfrac{2\,756}{100} + \dfrac{123}{99\,900} = \dfrac{917\,789}{33\,300}$
4. $\dfrac{5\,063}{3\,300}$

Convergence

Determine whether the following series converge. Justify your conclusion!

1. diverges by p-series test (Corollary 5.1)
2. diverges by the Comparison Test (Theorem 5.10)
3. converges by Limit Comparision Test (Theorem 5.11)
4. converges by the Comparision Test (Theorem 5.10)
5. diverges by the Divergence Test (Theorem 5.8)
6. Converges by the Ratio Test (Theorem 5.13, Theorem 5.6:8)

7. diverges by the Comparision Test (Theorem 5.10)
8. converges by the Limit Comparision Test (Theorem 5.11)
9. converge by the Comparision Test (Theorem 5.11)
10. converges by the Limit Comparision Test (Theorem 5.11)
11. diverges by the Comparision Test (Theorem 5.11)

Absolute and Conditional Convergence

Determine whether each infinite series converges absolutely, conditionally or diverges:

1. converges conditionally
2. diverges
3. converges absolutely
4. converges conditionally
5. converges conditionally
6. converges absolutely
7. converges conditionally
8. converges absolutely
9. diverges

5.3.3 Power Series

Interval of Convergence

Find the radius of convergence and the interval of convergence of the power series:

Radius of Convergence

Find the radius of convergence:

Limit Function

Find the sum of each power series and give its interval of convergence:

Expansions

Establish the following indenties:

Limit Function

Determine the limit function:

Taylor Series Expansions

Find the Taylor expansion at x_0 for each function and give its interval of convergence:

INDEX